William H. Calvin

THE THROWING MADONNA

Essays on the Brain

McGraw-Hill Book Company

New York St. Louis San Francisco Bogotá
Guatemala Hamburg Lisbon Madrid
Mexico Montreal Panama Paris San Juan
São Paulo Tokyo Toronto

ISBN 0-07-009665-1 {H.C.}
0-07-009664-3 {PBK.}

1 2 3 4 5 6 7 8 9 FGRFGR 8 7 6 5 4 3

Library of Congress Cataloging in Publication Data

Calvin, William H., 1939–
The throwing madonna.
Bibliography: p.
Includes index.
1. Brain—Addresses, essays, lectures.
2. Human behavior—Addresses, essays, lectures.
3. Neurolinguistics—Addresses, essays, lectures.
4. Women—Addresses, essays, lectures. I. Title.
[DNLM: 1. Brain—Physiology. 2. Evolution.
3. Dominance, Cerebral. 4. Neurons—Physiology.
WL 300 C168t]
QP376.C33 1983 153 83-884
ISBN 0-07-009665-1
0-07-009664-3 (pbk.)

Book design by Grace Markman.

For my mother,
AGNES LEEBRICK CALVIN

Contents

FIRST THINGS xv

Preface xvii

Traditionally served up as a *de rigueur* mélange of graceful acknowledgments with linked *mea culpa,* author's prejudices and *bon mots, caveat emptor* warnings to professional colleagues, and the usual succinct scintillating sentence summarizing the book, planted by the author in the hope that it will be picked up by an overworked book reviewer facing a deadline. Not to be missed.

ETHOLOGY AND EVOLUTION 1

1 | The Throwing Madonna 3

Tracing right-handedness back through the centuries from the invention of writing to how mothers carry babies on the left side. With a brief digression on pacifying babies, as practiced by pediatric neurologists, and how the left-sided sound of the maternal heartbeat quiets infants. Did right-handedness all start with a mother with pacified infant, throwing stones with her right hand to hunt rabbits?

2 | The Lovable Cat: Mimicry Strikes Again 14

As T. S. Eliot intoned, "A CAT IS NOT A DOG." As it is hardly a working animal, whatever caused the domestica-

tion of the cat? Perhaps it learned to mimic human babies, even though not looking anything like a baby? This chapter illustrates the "instinct" level of brain organization, what Konrad Lorenz called innate action patterns and the stimuli which trigger them. A tale of evolution, nonpoisonous snakes impersonating poisonous ones, and what prompts people to respond to a cuddly baby. The reader is prompted to perform a similar ethological analysis of the dog's success in pleasing people.

3 | Woman the Toolmaker? 22

The usual presumption is that the first human tool-users were male, but it certainly isn't turning out that way as ethologists study primates in their natural habitats: the starring roles usually seem to be played by females. Chimpanzees crack open nuts using rock hammers—and with considerable foresight and sophistication. But the males only engage in the simplest kinds of shell cracking, with females practicing the two more sophisticated techniques and flaking stones in the process. And even termite fishing, that classic example of chimpanzee toolmaking, turns out to be largely a female preoccupation.

4 | Did Throwing Stones Lead to Bigger Brains? 28

The hominid brain has enlarged threefold in the last few million years and, along the way, acquired specializations for language, handedness, and even music. What were the selection pressures which shaped this rapid evolution? An examination of one-handed throwing of stones at prey and how this cultural practice could have led to the enlargement of the brain—and provided the foundations for language cortex.

5 | The Ratchets of Social Evolution 43

A possible origin for empathy lies in the skills used for sizing up another person, as when bargaining in an oriental bazaar. The human brain has a special area for recognizing faces, even for evaluating the emotions revealed in another person's facial expressions. Which presumably proved useful in detecting when another was going to cheat rather than cooperate. But the evolution of cooperation itself is much harder to understand, because it requires that an animal forego immediate advantage for a chancy long-term gain. Now Axelrod and Hamilton have shown, using game theory, a simple "tit for tat" limited cooperation strategy that could have evolved even in a world of me-first types: it relies strongly upon recognizing others as individuals. Is this the origin of cooperation?

NEUROPHYSIOLOGY 53

6 | The Computer as Metaphor in Neurobiology 55

What will happen in some future year when an ancient computer is dug up by an archeologist, but the instruction manual is missing? An introduction to how neurobiologists investigate the brain and its hierarchies—and some cautions about "the physicist's fallacy."

7 | Last Year in Jerusalem 63

A real summit meeting of neurobiologists—atop Mount Scopus, with alternating views of Jerusalem and the Dead Sea. With expeditions to the mine fields of the Golan Heights in search of the elusive leech, to the depths of the Red Sea to see the molluscan "Spanish Dancer"—and then the true hazards (Thanksgiving viruses and Christmas bureaucracies).

8 | Computing Without Nerve Impulses 86

Over 99 percent of the nerve cells in the eyes reading this sentence are functioning without using nerve impulses. Yet the nerve impulse (a stereotyped voltage "spike" set off when a threshold voltage is exceeded, rather like pulling the trigger on a gun) has engrained our thinking about the way nerve cells compute for over a hundred years. The new view of the common currency of the nerve cell: variable volts, not unlike the variable balance of a bank account.

9 | Aplysia, the Hare of the Ocean 94

The secret of this seagoing slug's success is that it tastes bad—indeed, its major known predator is the neurobiologist, who prizes it for the knowledge its small brain yields about learning. A classic neurobiology story, analogous to studying small-town sociology where everyone knows one another rather than anonymous big-city masses. It explores how the interconnections between nerve cells change with learning.

10 | Left Brain, Right Brain: Science or the New Phrenology? 102

Does the vocal left brain dominate the poor artistic right brain? Or does the popularization of left-brain dominance involve the "worst kind of mixed metaphor, the kind that mixes up metaphor with reality"? An examination of lateralization and a modest proposal for a splashy new best-seller guaranteed to make the book clubs salivate: *Cooking on the Right Side of the Brain,* for the new holistic school of cooking which avoids mixing ingredients in sequential order (left brain, an obvious no-no) by dumping everything into the mixing bowl simultaneously.

WHEN THINGS GO WRONG 109

11 | What to Do About Tic Douloureux 111

Tic douloureux is a benign disorder producing excruciating lightninglike pains in the face. Yet unlike most chronic pains, it is almost always curable. Told as a medical detective story, this is a tale resplendent with the classic clues of the mystery thriller.

12 | The Woodrow Wilson Story 117

From dyslexic child to university professor, from university president to governor and then president of the United States. But between being a Princeton professor and the President who attempted to shape the Paris peace conference and the League of Nations, he suffered one stroke after another—and seemingly recovered from them all. But the gradual personality changes apparent before Paris were dramatically augmented by the influenza attack which interrupted the conference, and President Wilson was never the same thereafter. His major stroke five months later, while he attempted to persuade the Senate to ratify the treaty, left him not merely paralyzed on his left side, but unable to understand that he was disabled. Even if the Twenty-Fifth Amendment to the U.S. Constitution had been in effect then, it seems unlikely that the Cabinet could have declared him disabled—because Wilson would have fired them first, just as he actually fired his secretary of state for discussing the disability with the Cabinet.

13 | Thinking Clearly About Schizophrenia 127

One thing that schizophrenia isn't: it isn't split personalities like Dr. Jekyll and Mr. Hyde. It is a disease of young adults, largely inherited, which is often characterized by hallucinations which the sufferer may have difficulty distinguishing from reality. But until we understand nighttime dreams better, we are unlikely to fully appreciate how the brain normally maintains our grip on reality.

14 | Of Cancer Pain, Magic Bullets, and Humor 139

The evolutionary usefulness of pain following injury is examined. But evolution has not effectively protected us from useless pain: the pain of otherwise harmless neuralgias, the pain of cancer. A progress report on pain clinics, peptide hormones, research teams, and the sorry state of research funds.

NEUROLINGUISTICS 151

15 | Linguistics and the Brain's Buffer 153

Does language make use of the same "holding buffer" neural circuits as does the rapid motor sequencer evolved for rock throwing? When rearranging a complicated sentence into the deep linguistic structure emphasizing actor/action/object, are we making use of primitive throwing circuits? A short excursion into linguistics, that meeting ground of the humanities and the natural sciences.

16 | Probing Language Cortex: The Second Wave 163

Working from the bottom up, neurobiology focuses on membranes, nerve cells, circuits, modular collections of circuits exemplified by the hypercolumn—but then what next? Working from the top down, we can distinguish linguistics, instinct and memory, hemispheric organization, cortical maps, but then . . . ? How do modular super-hypercolumns generate grammar, store the word "rabbit," recognize a fuzzy animal as a rabbit, pronounce the word "rabbit"—or set about throwing a stone at a rabbit? What are the in-between levels of the hierarchy of the brain? An examination of the natural physiological subdivisions of language specializations of the human brain, together with male/female and IQ differences. The history of language physiology seems increasingly reductionistic as quite specialized patches of cerebral cortex are discovered, but they all surely work together in a committee—if we can only figure out how. And now there is some information on how they are orchestrated.

PROSPECTS 181

17 | The Creation Myth, Updated: A Scenario for Humankind 183

Every culture has its own creation myth relating the origins of humankind, invariably starring its own people, and the scientific culture is no exception. Neoteny, that phylogenetic trend which makes adults look more and more like juveniles, says something about possible genetic mechanisms. But the original problem, way back before brains started enlarging, was that there were not enough babies to comfortably maintain the population level. The solution to this problem of ecological economics involved upright posture and carrying, pair-bonding and nonprocreative sex. From this ancestor, somewhat different from modern-day great apes, the great brain boom began. A speculative scenario for our ancestors, but, unlike our usual creation myth with the unusually clever in the starring role, it suggests that our intelligence was an offshoot of a more mundane development in hunting and hammering.

Appendixes 197

Appended are two articles from scientific journals which

present material from chapters 1, 4, 16, and 17 in greater detail for the truly dedicated, insatiable reader. **Appendix A:** "Did Throwing Stones Shape Hominid Brain Evolution?," reprinted from *Ethology and Sociobiology.* **Appendix B:** "Timing Sequencers as a Foundation for Language," reprinted from *Behavioral and Brain Sciences.*

Bibliography 221

Literature citations, organized by chapter.

Index 241

FIRST THINGS

There is a vast untapped popular interest in the deepest scientific questions. For many people, the shoddily thought out doctrines of borderline science [parapsychology, astrology, ancient astronauts] are the closest approximation to comprehensible science readily available. The popularity of borderline science is a rebuke to the schools, the press, and commercial television for their sparse, unimaginative and ineffective efforts at science education, and to us scientists, for doing so little to popularize our subject.

Carl Sagan, Broca's Brain

. . . aversion to science is [surely] an unnatural state of mind. Science is the natural searching within ourselves and in our surrounds for explanations. It is the process of making comprehensible, by discovering and explaining with simple laws, that which had been dark—and often frightening—mystery. Science is curiosity in harness, and curiosity is surely one of the most elemental human motivating forces.

Mahlon B. Hoagland, The Roots of Life

New knowledge should ennoble not merely those who seek and find it, nor their immediate colleagues; it should add to the civility and wonder and the nobility of the common life.

J. Robert Oppenheimer, The Need for New Knowledge

Preface

Sometimes, an indirect approach is better. When you are stuck trying to remember something with a head-on approach ("Now, what *is* her name . . . ?"), it is often best to think instead about other, related things (perhaps the last time and place you met)—and hope that everything will fall into place. Tackling the mind-brain-intelligence troika in a head-on manner with verbal tools has been done traditionally by many philosophically inclined writers, such as Hofstadter and Dennett in recent years.

Neurobiologists are not uninterested in the big questions, but tend to feel that they too require a more indirect approach: thoroughly familiarizing oneself with the biological bases of animal and human behavior (ethology), with evolutionary processes and how they create stable hierarchies, with the bottom-up approaches to brain organization which start with membranes and neurons and circuits. And the top-down approaches which look at linguistics, instinct, hemispheric specializations, cortical maps, selective attention mechanisms, language physiology, and characteristic malfunctions.

Neurobiologists also have a broad practical knowledge about what works and what doesn't, ranging from the neurosurgeon trying to prevent a patient with epilepsy from hearing strange voices to the zoologist who learns to manipulate the caterpillar's nervous system as it changes over at metamorphosis from controlling crawling to controlling flying. And their tools are not merely verbal and mathematical: neurobiologists are always driving the medical instrument makers crazy with requests for tiny scissors smaller than the eye surgeons use, always chiding the computer manufacturers for making it hard to interface computers to animal brains.

But the head-on approaches from philosophy and artificial intelligence seldom raise my blood pressure. There are, however, some approaches from the direction of the physical sciences which I consider insidious due to their superficial reductionism. It's not merely that I've heard too many physicists and molecular biologists talking about decoding the brain as their next frontier, proposing to bring order to that computerlike enigma with their well-known mathematical reasoning. It's not merely that they, rather than the neurobiologists themselves, have been writing most of the popular books on neurobiology (a major exception, highly recommended, is *Programs of the Brain* by one of the most eloquent of all neurobiologists, J. Z. Young). Contrary to the grating tendency of some historians of science to treat biology as an imperfect stepchild of physics, many biologists originally trained in physics (the plop-plop sound and the oily smell of a vacuum pump, which characterize the corridors of physics departments, still stir fond memories for me) come to consider the methods of physics as simply inadequate for many aspects of biological hierarchies, especially evolution itself.

The reductionistic approach is, of course, a very important part of neurobiology (chapters 8, 9, and 16 show some classic examples), and a steep descent through a few layers of the hierarchy can indeed be exhilarating. But while digging deeper is all very well, there are too few people who attempt to understand the breadth of a particular level, such as comparing the various styles that neurons exhibit in making decisions. There seem to be several reasons. First, there is the competitiveness of researchers: we scientists like to see our names in print, and synthesizing never produces as many publications per year as tending a well-oiled digging machine. The second reason is more insidious because it is a form of self-definition growing out of the work ethic: just as narrow-minded economists define "work" so as to exclude what housewives do, many scientists define scientific productivity so as to exclude anything that isn't excavation labor. Depth counts, not breadth. And young researchers quickly learn that this narrow concept of productivity is what counts in getting promotions and grant money; they ignore it at their peril. (For example, writing a review article synthesizing several fields often doesn't count, but the tenth variation on a familiar excavation theme always counts. Of course, since most science researchers work for the "public sector," we all—by the economists' definition—exhibit zero productivity!)

This abused prestige of reductionism has had especially serious effects upon whole fields, such as our investigations into the origins of biological systems. People working in paleontology, developmental biology, neurobiology, and ethology have been essentially told to just report the facts and leave it to the population geneticists to explain how it all came about. As if discovering the evolutionary history within a particular level wouldn't be as interesting as the gene-jumping some layers below. Alas. That's like ignoring the development of sophisticated computer software in favor of reducing everything to quantum mechanical tunneling in the semiconductors from which the computers are constructed.

People have been driven to writing books themselves with even less provocation than provided by all of the bad examples. But the last straw was when a *Reader's Digest* editor phoned me, asking was it really true that, if a computer was built the size of the 10-billion-element human brain, it would consume more power than a New York skyscraper? As it happens, this is a figure extrapolated from those cited by the famous mathematician John von Neumann in his little book from the mid-fifties, *The Computer and The Brain,* still seemingly the major reference source for some physicists on matters neurological.

A few things have happened in the last quarter-century to both computers and neurobiology (my little word processor idles along at 150 watts, six times the basal metabolic rate of the poor brain which it assists, but the computer figure would have been another thousandfold higher back in von Neumann's time; as for neurobiology's progress, that just happens to be the topic of this book). Which left me trying to persuade *Reader's Digest* that there really were more interesting aspects to the human brain—if its power consumption of 25 watts is more efficient than a big computer, so what else is new . . . ? And repeating to *Reader's Digest* my standard refrain about the numbers game—that of those famous 10 billion neurons comprising the brain, 100 billion of them are located in just one section of the brain (the granule cell neurons of the cerebellum). Everyone's always underestimating the brain.

I did try tracking down that 10 billion rumor to its source, since even neurobiologists have been known to repeat it: 10 billion is really the estimate made by some quantitative Russian anatomists for the number of neurons in only the cerebral *cortex* of *one* hemisphere. Along the rumor mill, someone confused one hemisphere with two, then someone else confused the cerebral cortex with the whole brain. So there are 20 billion neurons total in the cerebral cortex. But the cerebral cortex, important as it seems to be in the brain of modern hominids, is only—to use a singularly appropriate metaphor—the frosting on the cake. The cake's neuronal numbers will have to be estimated by someone more brave than I. Besides, I have enough trouble just trying to count the cake's layers. Numbers, though of importance (see Chapter 4 and Appendix A), do not do it justice.

What evolutionary process created that frosting? It is indeed ironic that the term evolution (which literally means unfolding) is applied to the origins and development of the human cerebral cortex, which is noted for its extensive convolutions (deep folds in its surface which increase its surface area manyfold). Evolution by convolution—unfolding by infolding. Like an Escher print of a Möbius strip twisting around upon itself? Or perhaps a more appropriate image would be the deep folds revealed as a Georgia O'Keeffe flower opens up.

And it seems likely that science will need such artistic images if we are to succinctly summarize our understanding of the brain's evolution: just as the artist pares away the inessential detail and rearranges things so as to strike the viewer's eye with special impact, so the scientist tries to reduce things to essentials and present a representation so that the human mind can readily comprehend it. *But the essentials, for science as well as art, are not elementary particles but rather principles of organization, of composition.* Georges Braque said it well: "I do not believe in things, I believe only in their relationships." Our attempts to describe human brains are still the fledgling efforts of the child with watercolors, but we look forward to the various schools of brain art which may emerge.

Many thanks to my friends, relatives, editors, and colleagues who took the trouble to criticize the writing and the science. Simplification is a double-edged sword. It is, in many ways, the essence of science rather than (as some would have it) the imprecise refuge of popularizers. As Karl Popper recently said, "Science may be described as the art of systematic oversimplification—the art of discerning what we may with advantage omit." But I am also writing about neurobiology as the community endeavor of many people, so I am reminded of the other side of simplification, noted by a historian of molecular biology:

> Confronted with this living fabric unrolled abundantly upon his worktable, the historian picks up his shears, hesitates, and puts them down again. Whatever he does, he will be like the damned eighteenth-century Dutchman who made Rembrandt's *Night Watchman* smaller to fit a smaller room.
>
> Horace Freeland Judson,
> The Eighth Day of Creation, 1979

Though not a history of neurobiology, this book shares the same dilemma in selecting topics. While it touches on a spectrum from behavior to membranes, they are chosen because I can write easily about them, not because I have attempted to distill and evaluate in the manner of the historian.

I hope to persuade the reader that there is a new picture of human origins emerging from neurobiology and sociobiology which complements the more traditional methods of anthropology and linguistics. Some of what I present is new data, some is informed speculation, much is just bringing together traditional data from scattered sources.

It gives a picture of where science may be headed in its quest for human origins—but it is just one picture, mine, though shaped by many influences such as my colleagues' wide interests. This picture is likely flawed in some significant way—that's the (fortunately self-correcting) history of science, after all. I hope that the book whets appetites rather than satisfying them (to this end, I have provided many leads for further reading) and that some of the readers will become the thinkers and doers who will shape the future versions of the story.

Oh, yes—just for insurance, *mea culpa.* And *Shalom, achshav.*

W.H.C.
Seattle

ETHOLOGY AND EVOLUTION

He who understands baboon would do more toward metaphysics than Locke. *Charles Darwin*

Get used to thinking that there is nothing Nature loves so well as to change existing forms and to make new ones like them.

Marcus Aurelius
(*Antoninus, Roman emperor* A.D. *161–180*), *Meditations*

Natural selection is not always good, and depends (see Darwin) on many caprices of very foolish animals.

George Eliot

A 1982 Gallup poll of American adults revealed that:

9% agreed with "Man has developed over millions of years from less advanced forms of life. God had no part in this process."

38% agreed with "Man has developed over millions of years from less advanced forms of life, but God guided this process, including man's creation."

44% (a quarter of whom were college graduates) said they accepted the statement that "God created man pretty much in his present form at one time in the last 10,000 years."

9% had no opinion on creationism.

If we were specially created, then why were our genes created in the image of the African apes?
Vincent Sarich, commenting on the 99 percent identity of human and chimpanzee genetic material

1

The Throwing Madonna

Dux femina facti.
[A woman was the leader of the enterprise.] *Virgil, The Aeneid*

Their manner of writing is very peculiar, being neither from the left to the right, like the Europeans; nor from the right to the left, like the Arabians; nor from up to down, like the Chinese; nor from down to up, like the Cascagians, but aslant from one corner of the paper to the other, like ladies in England.
Jonathan Swift, Gulliver's Travels

Whenever a man stands up and proclaims himself as something special, unequaled in the world, we naturally become a little wary. Yet humankind as a whole has long been doing exactly that, proclaiming ourselves as special creatures for a long list of reasons.

There have, of course, been some rude setbacks to our self-esteem. The sun turned out not to revolve around us after all, but we got used to that. And then Darwin explained how one animal species evolves into a new type of animal—and did it so convincingly that many educated people have put aside their seemingly natural notions that humans were specially created in a manner quite unlike ordinary animals. There is something built into our brains which makes us curious about our origins. We make up stories and try them out for size. We (at least in science) eventually discard most of those stories in favor of other ones that fit the facts better.

Not, of course, that this has prevented the Arkansas state legislature from trying to make inconvenient facts go away—though, to put their recent promotion of biblical creationism in perspective, rumor has it that about fifty years ago, this same august body declared that, for the convenience of schoolchildren within Arkansas, the irrational number *pi* (3.14159 . . .) would henceforth be equal to exactly 3. (I have always imagined a capitol rotunda hanging over them, whose architect followed their rule in computing the dome's circumference from its diameter—and thus left a large crack.) Delusions about evolutionary biology have also been known to lead to catastrophe. Remember the "master race"? Or Lysenko's doctrinaire genetics and its disasters for Soviet agriculture?

The list of attributes that separate humans from our distant cousins, the great apes, has been shrinking as our study of animal behavior has intensified. Culture, toolmaking, language, "consciousness"—you name it, and some aspect of ethological research can probably provide an example from another animal. Instead of a quantum jump in ability separating us from the apes, we usually find that differences are more a matter of degree—that humans have developed some attributes (say, language-related attention span) to a greater extent.

Handedness, however, still survives on the human uniqueness list. No other animal species consistently prefers the same hand for certain skilled actions (well, let us say "vertebrate" so that we don't have to worry about the lobster or snapper shrimp with its one giant crushing claw). Individual rats and monkeys may prefer to use their right hand, but there are always enough other rats and monkeys preferring left hands for similar actions that it all averages out—the populations as a whole lack a side preference. So how did human handedness arise in evolution? And why *right*-handedness as the predominant preference rather than left-handedness—chance, or just another outcome of natural selection operating upon our ancestors?

If asked whether we are right-handed or left-handed, most of us would reply according to what hand we use for handwriting. Only 0.34 percent of us claim to be able to use either hand for writing a legible message. While 10.60 percent of us use the left hand, 89.06 percent write with the right. But if one tests other manual skills, the percentages are quite different. Ballistic skills, such as throwing a ball or swinging a hammer, are strongly right-handed, while fine motor skills without a rapid sequence of movements are considerably less specialized. For example, threading a needle is only 77 percent right-handed.

Surely some skills, such as throwing stones, are more primitive than others—say, threading needles. Indeed, our prototype of handwriting must itself be a secondary use of some more primitive aspect of handedness, having existed only since the invention of writing some 5000 years ago. Considering the literacy rate for most of those centuries, handwriting skills have been very lightly exposed to the selection pressures that shape genes in the Darwinian manner. Perhaps the head scribe could have supported more children than an illiterate blacksmith, but it seems a marginal argument at best. So one might expect the right/left aspects of reading and writing to provide us some good examples of how new lateralizations arise from old ones. Thus, the following warm-up exercise in historical ethology before the main event.

Written symbols (abstract, as opposed to cartoonlike pictograms) are thought to derive from the tokens used to keep tax records. A small bullet-shaped stone represented a bushel of grain, a long cylindrical stone stood for a herd animal, and so forth (the ancient Sumerians used at least eighteen such tokens in 3000 B.C., just before writing took off). The tokens were likely kept in jars by the tax collector and, like the coins in a modern piggy bank, had to be removed to be counted each time. Besides this disadvantage, there were probably instances of fraud, moving a few "goats" or "bushels" from the jar representing one person's tax payments over into another's jar.

Some Sumerian genius solved these problems by pressing the tokens into a slab of soft clay and then setting it out in the sun to dry and harden. The tokens probably didn't stick very well, but even if they fell out, the impression showed which token had been there. So soon the impressions of the tokens came to be used instead, merely "read" from a permanent tablet. And at some point, another genius just drew the shape of the token in the clay with a stylus, perhaps after breaking the last token of the desired shape. Thus, writing without typesetting evolved (the reinvention of movable type some 4500 years later was, of course, a major event).

So far, there is nothing in this somewhat fanciful reconstruction of events to suggest a right/left factor. The earliest tablets could probably be read from right to left just as easily as from left to right, since reading was just a matter of recognizing a token's shape and then counting them (mathematicians conjecture that this was the last stage of quantity counting which preceded the development of abstract numbers).

But, given the ease of forgery of clay tablets, someone tried chiseling the symbols into good hard rock (though, of course, some other side of human nature might equally well have provided the motivation—say, the need to aggrandize an "eternal" king by recording his wealth on a monument).

Our do-it-yourself ethology exercise even provides some real exercise for right-handed readers: Try holding a chisel and hammer (imaginary ones will do if real ones are not handy) up in front of you, chipping away delicately at the right side of a "tablet"—and then moving as far left as you can. Then try clipping left to right instead.

Done? If most of the ancients were equally right-handed, they too would have preferred to start from the easy side and work toward the more awkward side, rather than vice versa—and thus (so the story goes) the right-to-left sequence of such ancient written languages as Hebrew.

But most written languages scan from left to right; the usual rationale, again based on right-handedness, is that this avoids smearing the ink with the right hand, which might occur if working from right to left instead. Thus the switch from right-to-left to the predominant left-to-right ordering of symbols on a line, wherever the old traditions were not too firmly entrenched.

While such "just-so" stories suggest explanations for the right/left aspects of reading and writing, those aspects derive from the human predominant right-handedness—and force us to consider why hammering is easiest to do with the right hand for 88.24 percent of us. And whether hammering is itself just a secondary use of some more ancient skill originally shaped by Darwinian natural selection. The right-handedness of writing and hammering is exceeded only by swinging a racket (89.34 percent) and, strongest of all, by throwing a ball (89.47 percent). Both of the latter involve rapid overarm movements and both have suitably ancient counterparts (clubbing and stoning); indeed, hammering looks like a refined version of the very same shoulder and elbow motions employed in both clubbing and throwing. Might modern handwriting be a further refinement, employing the same neural machinery?

Neither clubbing nor throwing is necessarily a one-handed skill. Clubbing can often best be done using both hands and swinging the club over the head. But chimps certainly engage in one-handed hammering to crack open nuts (and, as we shall see in Chapter 3, with surprising precision). Chimpanzees throwing large rocks during threat displays often use both hands—but chimps also throw one-handed, both underhand and overarm, using the same motions as humans (though their aim is often almost random).

Our best prototype for early individual handedness is probably throwing because *advanced* throwing is distinctly one-handed—and because it has a sustained growth curve, with longer throwing distances repeatedly rewarded by natural selection in a manner hard to envisage for advanced hammering. The successful hunting of small mammals and birds (too fleet to chase, except via a good fast rock) is particularly important, as Chapter 4 elaborates, because it may have allowed our foraging ancestors to live in many new habitats with minimal forage—thus supporting a hominid population explosion.

But why might one-handed throwing develop a right-sided preference? Even if some individuals developed a right-handed habit for throwing, others would have settled on the left hand—so that throwing should distribute about fifty-fifty in the population. But it doesn't. To what can we attribute the modern 89.47 percent *species* preference for the right hand in rock throwing? Might it have been a random genetic decision? Maybe, but first we must try to relate it to some other right/left aspect of our world—which is, on the surface at least, quite symmetrical. But under the surface of our bodies lie some asymmetries, such as the unpaired internal organs. The right side's liver and the left side's spleen are silent and hidden, seemingly unrelated to hands and arms—but the heart's left-sided aspect does indeed show some promise. My colleague Joan Lockard and I have come up with some evolutionary musings about how the heart could be related to both individual handedness and species right-handedness. They are speculations rather than tentative conclusions, but they illustrate the nature of the problem and what a proper solution might entail.

Actually, the heart is pretty much located in the center of the chest, contrary to folklore—but the sound of the heartbeat is loudest on the left (this is because the more muscular left ventricle creates much higher blood pressures than the right ventricle, thus engendering more turbulence and valve-closing noise). So, just listening, one identifies the heart with the left side of the chest.

But who listens to the heart? Infants do. And hence our throwing madonna.

Infants cry and fuss much less if they are allowed to listen to a tape recording of a heart beating (indeed, this electronic pacifier is now used in some hospital nurseries). Since they had nine months to get used to that rhythm, perhaps they do not appreciate being weaned from it. Lacking the marvels of modern machinery, one can always just hold the infant against one's chest. And guess which side is better?

So it is hardly surprising that three out of every four mothers observed in shopping centers are carrying their infant with their left arm. This has been going on for centuries before shopping centers: a survey of madonna-with-child paintings in European art galleries (indeed, of over 400 such artworks from four cultures) showed left-armed infant carrying in 80 percent of the cases. And it may be a very, very old practice indeed.

It could be argued (perhaps this has occurred to you already) that left-sided infant carrying could just be a trivial consequence of right-handedness, to leave the right hand free for doing its thing. If that were true, then most fathers should carry infants on the left side too, since they are almost as right-handed as the mothers, on the average. But men seem as likely to carry an infant with the right arm as with the left; they exhibit no preference at all. And mothers who do not begin cradling their baby shortly after birth (separation due to illness or prematurity, etc.) do not, as a group, develop the strong left-armed preference; this not only adds weight to the weaning-from-the-heartbeat theory but suggests that perhaps the infant trains the mother to carry on the left, by crying more otherwise. If we cannot uncover a trivial explanation for the maternal left-armed infant carrying (and arm strength and cultural factors seem unlikely from recent research findings), perhaps we should concede it as primitive—and ask if right-handedness is instead a consequence of the "left-armedness" of maternal infant carrying.

But how might one-handed throwing (our best candidate for handedness thus far) be related to maternal left-armedness? Via a hungry mother. We tend to think of gatherer-hunter bands, back before agriculture, with everyone paired off and specialized in certain roles—the men hunting, and the women gathering while tending their noisy infants. But our penchant for dichotomy and optimal arrangements may mislead us. It seems likely that before things became that specialized, back a few million years ago before the brain started its rapid enlargement, women also hunted. Indeed, they still do in some primitive tribes, mostly small game incidental to gathering. And, contrary to our fixation upon "big game" and cooperative hunting enterprises, it may have all started with small game and individuals.

One requirement would be to keep an infant quiet while stalking game. Distraction ("see the rabbit?"), or holding the infant up to smile at it (called *en face* in the literature), would hardly seem appropriate to the situation, besides requiring both parental hands. But newborns can also be temporarily quieted by such maneuvers as a nipple in the mouth (or a substitute; even a fold of blanket will serve if the infant is not too hungry), crossing the infant's arms across the chest and pressing them (this is said to work because the posture creates memories of the womb!), and wrapping the infant securely in a blanket (again, back to the womb). One-armed cradling of the infant beneath a breast borrows from all of these maneuvers (though one must be careful not to press on the infant's face while shadowing its eyes; that sets off a struggling, noisy reflex which serves to protect breathing space). None of these maneuvers is intrinsically left- or right-handed. But heartbeat pacification is.

Now, chasing an animal is even more tiresome than usual if you are carrying along an infant. It would certainly encourage a mother to practice throwing stones at nearby rabbits and birds. But one mustn't scare the prey away, and so women hunters would have tried to keep their infants quiet. And there is an advantage to cradling them against the left chest. So it would usually have been right-handed throwing that was most successful. Q.E.D.

But, you may say, surely this is a neo-Lamarckian argument (mother's acquired right-handed skills being somehow passed on genetically to a subsequent baby, a possibility considered quite unlikely on genetic grounds). No, there is indeed a standard Darwinian explanation for how natural selection could have shaped a predominantly right-handed population.

One side of the brain usually has better neural machinery than the other for orchestrating rapid movement sequences (other than locomotion, as has become apparent in the physiological study of some foundations of language skills). If a hunter's left brain was the better rapid sequencer, the right-handed throws should usually be faster (and thus go farther) than left-handed throws. If it was the right brain which was the faster sequencer, then left-handed throwing (and hammering, etc.) would be better.

Who would be the more successful hunters, the left- or the right-handed throwers? For hominid men not encumbered with infants, there might be no consistent side preference when averaging across the population. But mothers with left-brain sequencers should be better hunters (faster throws and quieter infants) than those mothers who had to hold their infants on the right side in order to use their best throwing arm. More of the infants carried by right-handed mothers (and in turn often carrying their right-handed genes) would survive than those of left-handed mothers.

This argument does not assume a female predisposition for left-armed infant carrying (though that may well exist too; as described in Chapter 5, facial emotions are best judged when they are seen with the left visual field); as noted earlier, the infant may merely train the mother to carry nearer the sound of her heart, by crying otherwise. So the only innate behavior assumed is that, in effect, the infant doesn't like being weaned from the sound of the heartbeat to which it became accustomed *in utero*.

Thus crying babies would cause more right-handed genes—really left-brain movement sequencing genes—to survive into subsequent generations. Like compound interest, even small differences grow when exposed to continuous selection pressures (punctuated evolution, rather than Darwinian gradualism, only means that such selection forces are most effective when acting upon a small, isolated inbreeding population about the time a new species evolves). Given at least 100,000 generations in the last several million years, much but not all of today's population could have left-brain rapid sequencing—approximately what we today call right-handedness in 89 percent of the human population? The evidence for predominant right-handedness goes back at least a half-million years to the grips on the flaked tools found with *Homo erectus* in Zhoukoudian, China.

Some such mechanism is needed to explain how handedness got started, how brain enlargement occurred, how lateralizations of functions came about. The human brain is distinguished from the brains of the great apes by an extraordinary extent of lateralization of function. *Language on the left, visual-spatial on the right* is an easy-to-remember dichotomy which hardly does justice to the mosaic of lateralized functions, but does serve to remind us just how much humans have departed from the usual primate arrangement of homologous hemispheres and duplicated "head offices" for most functions (though there are hints here and there of various asymmetries: For example, rat brains have thicker cortex on the right side). And rapid motor sequencing is a convenient candidate for an early lateralization because it seems like a good foundation on which many language specializations could be constructed, as discussed in chapters 4 and 16.

Our common heritage, shared with the great apes, may turn out to include several of the essential ingredients: a tendency of one hemisphere to be better than the other for rapid motor sequencing and even the practice (whether innate or acquired) of left-armed infant carrying. But unless mothers used that neural sequencing machinery for an important reason during infant carrying, there might be no environmental selection for left-brain sequencing. And thus (as we shall explore in Chapter 4) perhaps no selection for bigger brains and language. Did the great apes miss the bigger-brain bandwagon because mothers don't throw?

Yet one cannot draw such conclusions without a lot of detailed facts about primate behavior and motor systems neurophysiology. Primate ethology is a kind of behavioral archeology. To quote George Eliot again,

> Our deeds will travel with us
> from afar
> And what we have been makes us
> what we are.

And you don't have to visit a hot, dusty dig to see this kind of archeology in action: just look around, the next time you visit the zoo, for someone with a clipboard and a stopwatch who is intensely studying the mannerisms and play of the monkeys and apes—and perhaps comparing them to the visitors, too. Things like handedness and infant-carrying preferences are among our major clues to a smaller hominid brain that no longer exists, and careful study of them reveals far more than the casual glance and the cocktail party truism.

While this maternal hunting possibility for right-handedness is perhaps an improvement on its predecessor (an anatomically marginal hypothesis which few believed: a warrior holding his shield in his left hand to better "protect his heart"), most hypotheses turn out to be wrong or only part of the truth. At their best, such hypotheses inspire and provide a new focus for research: into why there is left-armed infant carrying, into why right-handedness is stronger for ballistic skills than fine motor skills, into the corresponding abilities of the great apes (for example, the relation of chimp throwing abilities to their better-developed hammering skills discussed in Chapter 3), into the neurophysiology of throwing and other ballistic movements.

So, caution. Yet, maternal-infant hypotheses have an inherent strength when arguing natural selection: for example, Lovejoy's theory that the upright posture allowed mothers to forage for food while carrying infants who couldn't cling to parental body hair in monkey/ape fashion. And skill in mothering is under very strong selection due to the often high infant mortality rate. Jane Goodall found that inexperienced chimpanzee mothers lost over half of their infants though experienced mothers lost only 17 percent (surviving human childhood became the norm only in the present century, and only in the developed countries). While silent hunting skills may be an important survival skill for mothers themselves, the infant is even more exposed to selection than the mother: the survival of the infant is threatened by even minor diminution in the mother's health or fitness (absent the backup provided by kinship and/or societal care).

Infant pacification and maternal hunger would seem a likely part of the evolutionary history of handedness—but whether this fascinating aspect of natural selection will prove to be the key factor, only more work will tell. Still, consider the possibilities. Today, only the portraits of twentieth-century male pitchers hang in the Baseball Hall of Fame. Yet perhaps someday—just possibly—a "Museum of Human Origins" might have its entrance hall dominated by a large painting depicting a turning point in hominid evolution: showing a mother with infant, a rock in flight, and a rabbit. Its title might read: *The Throwing Madonna.*

2

The Lovable Cat: Mimicry Strikes Again

My cat may manipulate me psychologically, but he'll never type or play the piano. *Stephen Jay Gould, The Panda's Thumb*

Our cat is called "Cat," though that might have horrified T. S. Eliot, who said that each cat must have three different names. But at least our previous cat, Macavity, was named after a famous poem from Eliot's *Old Possum's Book of Practical Cats,* the original cat book. Cat has endeared herself to us, despite her lack of understanding of the human habit of reading (she prefers to sit on one's chest, between eyes and book; our Macavity at least consented to sit on one's shoulder).

Given an evolutionary turn of mind, one cannot help wondering why cats can be so endearing. Perhaps they are the original Teddy Bear, taking advantage of the softness which human beings display toward human infants? Taking advantage of an emotional response is an old and honorable technique in the animal kingdom, though usually involving "protective mimicry," where predators avoid certain prey because they "taste bad." For example, Henry Bates noted in 1862 that some Brazilian butterflies have adopted the color and markings of poisonous species, but without the poison (rather like posting burglar alarm signs on your windows but not bothering to buy an alarm system). Mimicry is a fascinating trait in nature, demonstrating that natural selection sometimes operates on appearances rather than reality, that it isn't as specific as you might think, that indiscriminate "prejudice" is an important fact of life. All of which provides some clues to how the domestic cat became domesticated.

T. S. Eliot intoned: "A CAT IS NOT A DOG." One's evolutionary suspicions are not aroused so easily by the domestic dog. After all, dogs can be working animals, the surrogate shepherd or the hauling husky. Cats are sometimes depicted in humor as working animals: I've always liked the story about the cat that ate cheese so that he could breathe down the mouse hole with baited breath. The usefulness of cats is, however, borderline at best. While Cat proudly presents us with shrews, mice, and the occasional bird (usually in the middle of the night), would anyone take the trouble to domesticate a wild cat for such a purpose? However useful farmers may find cats around the barn (and cats were valued for catching rats during the plague years), something else surely was responsible for the domestication of the cat from its wild ancestor, the Kaffir (or Cape) cat, *Felis lybica* (which still is found on the Mediterranean islands, in North Africa, and even in Indian deserts).

One suspects that a cat adopts a human in the hopes of being adopted back—and thus fed. Certainly present-day cats play up to humans in a variety of ways. Consider the greeting ceremony. I am not thinking here of our cat's "Here I am, everyone" vocalizations upon entering the house, but of a feline behavior that few people realize is a greeting. It turns out that the characteristic rubbing up against furniture is mostly a greeting to people in the room; it doesn't often occur when a human observer is absent and a TV camera substituted. It is difficult to understand the function of this poorly appreciated greeting. Perhaps it is just another way of attracting attention? (Certainly when the cat rubs against me instead and I trip over it, it attracts my attention.) Or is it just a feline superstition? But the cat's lap-loving behavior is more easily understood and may well be the key to their success as a species: it evokes far more than mere attention.

One has only to observe a human holding a cat to realize what is going on: the pet is evoking the same reactions that a cuddly baby sets off. Their contented responses when cradled set off the same flood of emotion in us. Babies babble and nuzzle, cats purr and rub. Their not-quite-speech vocalizations cause us to respond with smiles and encouragement. They like being handled, and we like handling them. We like seeing them happy. There is something congenitally comfortable about cuddling.

(Cuddliness, defined? There is actually a rather pragmatic medical definition of cuddliness. Part of the neurological examination of a newborn infant is to cradle the baby and watch—and feel—to see if it begins to conform to the shape of the examiner's body, to nestle in. Babies with depressed function of the higher brain centers often fail to mold themselves in this fashion. Just try cradling a wide-eyed, alert cat that wants to be elsewhere: it too will seem hard and unyielding, even if not struggling to get down. Because newborns cannot talk and follow commands, there are few simple ways of testing their cortical functions—among the others is the response to pacification maneuvers noted in Chapter 1—and consistent lack of response on such tests can be a tip-off to the physician that more definitive tests may be in order.)

Of course, a cat is not a baby. It doesn't even look like one, and babies certainly lack whiskers. So this is not the usual sort of protective mimicry based on appearances. Predatory birds learn to avoid snakes with the characteristic colored bands that adorn the coral snake; the birds presumably think that such snakes "taste bitter." Ingesting venom is almost as bad as getting bitten, so birds without the ability to avoid coral snakes may not reproduce very well. Other snakes with similar markings, poisonous or not, seem to be avoided by the birds as well—so they survive better and produce more little snakes with similar colors. The birds presumably have enough other prey to eat that they haven't made their color prejudice a little more discriminating. This Batesian mimicry is just another elegant example of natural selection, of the survival of those species with a good gimmick or an imperfect predator. Is the cat still another successful mimic, despite the lack of physical resemblance to babies? Using sweetness rather than bitterness? Playing on human psychological propensities rather than physical appearances?

The evolution of *Felis domesticus* (also known as *Felis catus*) tells us something about the evolution of human reinforcers, about what we find sweet and pleasurable, and why. Although "sweet" has a specific meaning when looking at the neurophysiology of the tongue's taste buds, the concept of sweetness extends to other areas beyond food. There is a symbiotic relation between parent and offspring: the infant needs much assistance to survive, and the parent "needs" to propagate its genes. Genes which reinforce this symbiosis are to the benefit of both parties (and hence propagate the genes). Parents who respond to the sight of an infant by picking it up and cradling it are more likely to propagate their genes for doing so. Sweet is more than a quick-energy food related to our ancestors' fondness for fruit trees: it is usually a selfish gene at work in a more global way.

After all, nurturing an infant is not a universal trait: some species invest nothing in their offspring after birth. They just try for large numbers and leave them to fend for themselves (e.g., many fish and frog species). Other species invest a lot in a few offspring, the so-called "K selection." They had to evolve ties which bind parent and infant together psychologically for an appropriate period of time. A gene leading to cooing can interact with a gene leading to cradling, to the benefit of both. Indeed, any of our elementary human pleasures probably has such a background. Anything that is sweet today is a clue to what was good for the species a long time ago.

But the features which trigger such responses (Konrad Lorenz called them "innate releasing factors") need not be a perfect imitation of the original object. While the selection process may have involved a human baby, the trigger features may simply be anything about the right size which combines cradling and cooing (like the predatory birds and their snakes, nothing may have made us more discriminating in whom or what we cuddle). If the cat's purr can substitute for the coo, the cat has lucked in to a good deal. Because it can become the recipient of some of the affection usually reserved for babies, it will stand a much better chance of receiving food and shelter. Cats with purring genes, or genes leading them to seek being cradled, will survive much better than those without. Over a long period of time, wild cats in contact with humans will slowly develop some of those attributes which we humans find sweet, simply because we don't feel as compelled to feed and shelter the ones that are aloof. It probably started with orphaned kittens of *Felis lybica* being raised by some children playing house.

This "lap-first" theory says that the cat was domesticated by mimicry of a trigger feature of the parent-infant symbiotic relationship. And what are those trigger features? Certainly not whiskers! But the ethologists studying ducks haven't had to use stuffed ducks—something the right size and color, and maybe a little fuzzy, will work just fine to set off many behaviors in a real duck (indeed, for some ducks, Konrad Lorenz himself sufficed). For humans, one way to find out would be to use a stuffed animal and then subtract features until the cuddling response is lost. A do-it-yourself ethology assignment: Borrow some well-used stuffed animals with missing eyes, ears, or legs. Get some new versions, so that you have a full range of physical features. Now get a stopwatch and see how long people will hold each stuffed animal before laying it back down and selecting another. Soon you'll have a reasonable idea of the essential trigger features for this kind of sweetness. Alternatively, consult the doll manufacturers. Or the literature: Lorenz noted in 1950 that the animals for which humans feel affection have large eyes, bulging foreheads, chins that tuck in rather than jut out, and a springy elastic consistency—all characteristic of human infants.

They say travel is broadening, and this comfortable (cuddly?) view of cat evolution has been challenged by several encounters. The first involved a feral cat, a domestic cat gone wild. The Friday Harbor Labs are on a large nature reserve up in Washington's San Juan Islands near Victoria, British Columbia. The apartments for the researchers are in a sylvan setting: It's really a wonderland, where deer graze outside your window, where the rabbits don't even bother to interrupt mowing the grass when you walk past. At night, there is a parade of raccoons making the rounds of doors where they are likely to get handouts. Some are quite aggressive and will march indoors and stridently demand the food off your dinner table: real masked bandits. Others will hang back timidly.

One night in the midst of this parade came a most impressive cat. It had the kind of large, bright, all-seeing eyes that one tends to associate with human geniuses. It was quite wild, in the sense that it would not approach humans. There was clearly a working understanding between this feral cat and the raccoons; the cat may have been outnumbered, but one sensed that it had won a few arguments. Now they shared the bounty. But unlike the aggressive raccoons, the cat wouldn't come anywhere near the door. Timid may have described the raccoons that held back, but one would hardly label this cat timid: it lived and breathed intelligent caution. It wasn't skittish; it was crafty and dignified, engaged in a war of wits with two other species simultaneously. Was this feral cat acting like the wild predecessor of the domestic cat? Was the cat originally a camp follower, staying on the outskirts of human habitation and living off the garbage pile, gradually becoming tame? Was the lap the final step, rather than the first step? Not quite like wild kittens being raised as infant surrogates.

Later, during a sabbatical year in Jerusalem, came a massive exposure to "feral cats." The streets there were full of cats raiding garbage cans. And they were quite wary, never investigating an outstretched hand, much less a lap. I rather pride myself on my ability to make friends with any cat, but I was rebuffed every time. But the reason seemed rather clear: children threw rocks at them whenever possible. Adults cursed them—and not just the ones in competition for the same resource, such as the tramps who came around an hour in advance of the garbage truck, searching through the trash cans for anything edible or useful. Everyone seemed to treat cats as a pest, to be driven away.

Had the symbiotic relationship between cat and man completely broken down? And, since one tends to think of the people there as being a bit closer to the land and the conditions under which the original domestication of the cat might have taken place, what did that say for my pet theory for cat domestication through infant mimicry?

Might Israeli cats be a different subspecies than the North American cats, since they acted so differently and since no one seemed to treat them as pets? Most of the cats in North America are descendants of ones brought over from England or the Netherlands in the seventeenth century. Maybe, I surmised, the Jerusalem cats were closer to the cats that the Egyptians had worshiped; indeed, some of our earliest knowledge of cats comes from mummified cats dating back to at least 2500 B.C. Over 300,000 mummified cats were found a century ago; the study of their remains might have taught us much about the evolution of *Felis lybica* into *Felis domesticus*—but these feline mummies no longer exist, as 19 tons of their remains were shipped back to England by some entrepreneur to be ground up for fertilizer. Alas. Lacking data, one can always speculate.

Armed with my tentative theories about Middle Eastern cats, I consulted with an Israeli zoologist. She said that my data base was insufficient for another reason—that I just hadn't been in an Israeli home yet that had a pet cat. Many Israeli homes turn out to have pet cats. Some of those cats on the streets are indeed strays, but many are also someone's pet. They act entirely differently indoors, just like normal cats, seeking out one's lap. But outdoors, they've learned to run at the sight of people. From which one might conclude that the North American cats have simply been more successful at making friends—at adopting people, as it were.

So was it lap-first or lap-last; did people adopt the cat or vice versa? Having had a whole family group of hominid fossils dug up in recent years, perhaps we will just have to hope that a suitably ancient hominid will be found with a cat skeleton cradled in its lap.

The evolutionary analysis of the domestic dog, *Canis familiaris*, and the corresponding trigger features in human behaviors will be left as another exercise for the reader. (*HINT:* Dogs like to follow one around, gaze up admiringly, play catch, romp, be a good pal—all rather like a ______. The fundamental difference between "cat-people" and "dog-lovers" is therefore ______.)

Postscript

As Harold Morowitz points out, it is axiomatic in the publishing world that essay books don't sell, especially books of science essays. Lewis Thomas's *The Lives of a Cell* made best-seller status, however, one hopes confounding whoever invents such axioms. The obvious solution is to mimic a successful strategy, just as the cat mimics the baby's strategy for getting attention and affection. Thus one must include an essay mimicking either (1) a cookbook, (2) a sex manual, or (3) a cat book. The perceptive reader of this essay should be able to figure out the reasons for this particular troika leading the nonfiction best-seller lists.

Mimicking a cookbook might be fattening. Sex manuals require expensive illustrations (whose costs the publishers take out of the author's royalties, perhaps explaining why so few books, of other types, are illustrated).

But a cat essay only requires contemplating this endearing little monster on my lap who periodically tries walking across the keyboard of the word processor. It's not that I mind the random letters tacked onto my manuscripts—and there are often a lot, as the keys automatically repeat like a machine gun if held down. But sometimes Cat will instead hold down the backspace-delete key, causing my recent prose to be consumed backward from the end with great rapidity until I snatch her off the keys. I sometimes have to eat my words, but I prefer that the cat and computer not do it for me. In addition to her aforementioned failings regarding reading, Cat apparently doesn't understand writing either—but she instinctively understands human psychology. I haven't tried her out on a piano yet, but given her typing performance, I suspect Gould is right on all counts.

3

Woman the Toolmaker?

To the first flaker of flints who forgot his dinner. *W. H. Auden*

[Apes] are capable of perceiving the solution of a visible problem, and occasionally of improvising a tool to meet a given situation; but to conceive the idea of shaping a stone or stick for use in an imagined future eventuality is beyond the mental capacity of any known apes. *Kenneth P. Oakley, Man the Tool-maker* (1964)

Because modern big-game hunters are male, it has long been assumed that the earliest hunters were also male—an assumption examined in an earlier chapter. A similar assumption has the creativity of early hominid males resulting in toolmaking, such as the sharpening of stones for their spears. But, as more time has been spent in studying primates in their natural habitats, many examples of invention and tool use have been observed. And the females usually play the starring roles.

It all started in 1953 when scientists began studying a troop of Japanese macaque monkeys on the island of Koshima. Sweet potatoes were left out on the beach to attract the monkeys to a location where the scientists could observe them. An 18-month-old female, named Imo by the observers, adopted a novel method for cleaning off the sand adhering to the surface: she would dip the potato into the ocean and wash the sand off. Soon Imo's playmates, and then their mothers, adopted the technique. Within several years

90 percent of the monkeys in the troop had adopted this cultural practice, with only the old males refusing to have anything to do with it. The scientists tried introducing caramels wrapped in paper. The candies were first sampled by the young, then by the more permissive mothers, and finally (after about three years) most of the adult males had also adopted the new food.

In 1955, Imo again made a major invention. The scientists had scattered wheat on the beach. Picking it out of the sand, grain by grain, is rather tiresome. Some species of monkeys (such as the gelada baboons) are good at such tasks because they make their living by seed foraging in the plains; they also have evolved short forefingers to make it easier to pinch a grain between thumb and forefinger. Imo found a better way: She threw a handful of beach into the ocean (perhaps in a fit of frustration?). The sand sank, but the wheat grains floated. Imo scooped up the wet wheat and ate it—and thereafter repeated her performance whenever there was wheat available. Again the practice was observed and imitated by others, the juveniles being the most likely to adopt this new sifting method, the older animals being most likely never to learn the technique (they didn't mind the wet wheat, however, as they would plunder the juveniles', though never throwing it into the ocean themselves).

This study not only illustrates innovative cultural practices in another primate, but it also shows who is most likely to imitate a new invention: the young. In general, the young are likely to be more inventive as well, being more playful (Imo was also perhaps young enough not to share the usual primate fear of water). Why might one sex be better at invention and tools than the other? Because mothers are around the young more than the males, they may be exposed more to new inventions. But other factors seem likely to be involved in female inventiveness. Perhaps they have more patience with a difficult technique (as we laboratory-bound scientists are only too aware, nothing works the first time—or if it does, it fails the next dozen times). Or perhaps the females just have inherited more relevant skills. None of these possibilities have been sorted out yet, but the natural behavior is telling some pointed stories.

In the late fifties, when Jane Goodall began studying the wild chimpanzees on the Gombe Stream preserve in Tanzania, she noted that chimps would spend hours fishing for termites, especially the female chimps. This quiet, patient behavior was quite in contrast to the usual noisy, rambunctious assemblage of chimpanzees. A chimp would select a stick or twig of the correct thickness and length, strip the leaves off, and poke it down a hole in the termite nest. By withdrawing it slowly, a few termites might come along for the ride—and promptly be licked off. Chimpanzees have been observed to strip such a twig, without a termite hill in sight, and then carry the prepared tool around for some time while searching for a suitable termite hill in which to use it, contrary to the axiom that only humans prepare tools in advance of need. Subsequent studies by William McGrew at Gombe Stream have shown that female chimps are much more persistent at termite fishing than males, though males start fishing about as often.

The rest of the chimp story concerns nut cracking. Breaking open shells is practiced by many species. Birds, such as the Pacific gull, pick up shellfish and fly to a suitable height, dropping the shell over a rocky area (gravity as a tool, yet), and then descend to consume the innards. An interesting reversal occurs when the shell is too big to be carried aloft. In 1850, long before aerial warfare began, natives in South-West Africa told of how Egyptian vultures cracked open giant ostrich eggs by bombing them with rocks carried aloft in their talons. More recently, Jane Goddall and Hugo van Lawick have photographed this vulture actually throwing rocks at ostrich eggs: standing at short range, the vulture holds a rock in its beak and throws it with a strong downward movement of head and neck. But chimps take the prize: their technique for nut cracking is strikingly human.

In 1843, J. S. Savage and J. Wyman noted that chimpanzees may use rocks to hammer upon shells, in an effort to crack them open. But subsequent observations of such hammering have been rare. In 1979, Christophe and Hedwige Boesch of the University of Zurich embarked on a three-year study of the chimpanzees in the Tai National Park of the Ivory Coast (the neighboring tribes fortunately consider the chimpanzee sacred, an attitude which helps the chimps survive; if the American bald eagle were protected by religion rather than legislation, it would be better off too).

The Boesches concentrated on how chimps hammered open nuts—a frustrating task, because the chimps would run whenever they caught sight of the scientists. Out of 4200 hours in the field, the Boesches observed nut cracking for only 62 hours total. Often they would only be able to hear the nut cracking take place, staying at a distance and counting the hammer blows, learning to identify the sounds with the actual performance on those occasions when they could get close enough to see through the dense rain forest. When the animal left the site, satiated with nuts or scared off, they would be able to observe the chimp's sex and age.

Chimps gather up nuts and then find a hard surface on which to place them, such as a rock or the root of a tree. They then proceed to hammer them open, using whatever rock or natural wooden club is handy. The nuts on the ground are quickly consumed, so the chimps climb the tree to collect more. The tree, however, is at least 15 meters high—the height of a 4- to 5-story building—and it is hard work just to bring back down a few nuts. The chimps reappear hauling all the nuts they can carry in a free hand and inside their mouths. Then they crack the Coula nuts open on the ground in the usual manner, again using whatever natural hammer is convenient. And then the long climb back up the tree for more.

The hammering technique requires positioning the nut on the hard surface at the correct angle and then hitting it repeated times: first hard, and then softer and more precisely. Somewhat more females were observed using this simple gathering and hammering technique than males, but it was widely practiced by both sexes, by adults and adolescents, even by juveniles (and by one female infant!).

The Panda nut is much bigger and tougher, about 5 or 6 cm long, formidable like a giant walnut. It has three or four almonds, but each is surrounded by a hard shell and sticky husk, plus the thick outer shell. They are gathered and carried by hand to the nut-cracking workplace. Chimps place them on hard rock outcrops or in depressions in a tree root (probably created over the years by generations of chimps visiting the same tree and needing a good anvil). Sticks are too soft to be good hammers; only rocks are used.

If no rock is lying nearby, the chimps will search for a suitable one, sometimes carrying it hundreds of meters back to the workplace. Powerful hits are required to initially crack the thick outer shell of the Panda nut; then a series of hits are precisely graded to crack the inner shells without shattering the almonds. The nut must be repositioned at least three times during this process.

Over 95 percent of the chimpanzees observed to engage in this elaborate cracking process for the Panda nuts were females. And females were also virtually the only chimps to engage in the most skillful nut-cracking process of all: with the workplace high up in the tree, balancing precariously on a tree branch while hammering away at Coula nuts.

Saving the round trip up and down the tall tree is all very well, but it does require some advance planning: a suitable wooden club or rock hammer must be selected and carried up the tree using one hand or foot. The chimp first gathers some nuts, storing them inside the mouth and carrying by hand as many as possible to a suitable workplace high in the tree on a broad branch. To crack the nut, one hand must be free for the hammer, and if the nut is to be kept from falling to the ground after the hammer blow, the other hand often must hold it. This means that the chimp has to rearrange the stockpile of nuts, holding them with one foot (this leaves one foot free for hanging on!) and inside the mouth. After the target Coula nut is cracked by repeated blows and repositionings, it is eaten—which of course requires that the stored nuts inside the mouth must first be transferred to another holding place. And then there is the hammer to worry about: it is often balanced on the tree branch temporarily. Finally, the chimp gets to eat its hard-earned almond.

To hammer another nut, hammer and nuts are rearranged and the virtuoso performance repeated. When the working supply of nuts is exhausted, the hammer is sometimes left balanced at the branch workplace; other times, the chimp carries it along on the next round of gathering.

About 92 percent of the chimps using the treetop workplace technique were females. And the only two adult males observed to use the technique are perhaps special cases: they were in the same tree as an estrus female and probably followed her there (when a female chimpanzee is sexually receptive, she is often followed around for days by several adult males). *So the two most difficult nut-cracking techniques are practiced almost entirely by female chimpanzees. And the less difficult termite-fishing technique is practiced largely by females.*

Humans might deal with such nut gathering somewhat differently. If there were sufficient cooperation developed among the group, one member could climb the tree, shake loose some nuts, and then return to share the supply gathered off the ground by the other members. Or, lacking such trust, the human might use a carrying bag slung over one shoulder, picking a large supply in the tree and then returning to earth to find a less precarious anvil. The lack of either cooperation or carrying bag meant that these chimps developed treetop hammering skills that had the precision not to knock the nut off the limb or smash the fingers holding the target nut.

Most speculation about the development of toolmaking has centered upon its relationship to hunting, though some tools (such as pry bars and hand axes) seem more likely to have arisen from gathering activities (digging up edible roots and grubs). In particular, precision hammering is usually associated with flaking rocks in order to make scrapers or hand axes, or to obtain flakes for sickles or arrowheads (flaked pebbles, probably for use as knives, have been found in deposits several million years old). But the chimp hammering techniques suggest that rock flaking itself could be an accidental by-product of nut cracking: the Boesches observed that chimps sometimes flaked stones while pounding the hard Panda nut. In three instances this happened to a granite rock and once to a quartzite one. And in each case, it was a female chimp who flaked the rocks.

The data on the evolutionary origins of toolmaking are still thin, but the shoe is now on the other foot as regards which sex was the more innovative. The Boesches conclude their discussion of the flaked stones: "these chimpanzee-made artifacts suggest that such tools could have been produced by early hominids when they used stones as hammers during a gathering activity. . . . The skill of female chimpanzees at Tai suggests the possibility that the first human tool-makers were women."

4

Did Throwing Stones Lead to Bigger Brains?

Modern evolutionists concentrate on what they can see and measure. This is good, but only up to a point. Some components of evolution that cannot yet be measured have probably been important too, and throwing may be one of them.

Philip J. Darlington, Jr., *Evolution for Naturalists,* 1980.

What is now proved was once only imagined. *William Blake*

Friday Harbor Laboratories
San Juan Island, Washington

Lying on the beach and throwing stones at a log seems a natural sequel to a really low tide. Low tides mean hopping from one slippery rock to another as the waves gently lap the exposed seaweed, then bending over and peering under the rocks in search of intertidal creatures. Such unaccustomed exertion in the fresh salt air is best followed by finding a sandy stretch of beach and settling down with binoculars, or a good book, or a good pillow. For a break, throw a stone at a log.

The knot in the driftwood log makes a good target. Throwing stones is such an unexpected pleasure. It brings back memories of three children having a contest to see who will be the first to topple a pile of flat stones placed atop a distant log. And it isn't just children (and similarly playful types such as scientists) who like to throw rocks at targets. What is it that makes slinging a stone such an elementary pleasure? It almost seems in a class with responding to a baby's smile, or taking pride in one's handiwork—as if it were one of those built-in human reinforcers. A pleasure that happens to have survival value for the species.

That seems to be one of the things that have changed in recent evolution: while young monkeys playing together romp unmistakably like children, no one ever sees monkeys lobbing rocks at a log and cheering one another on. They do use projectiles: monkeys have been observed to drop coconuts from trees to crack them open. But then birds do that too, breaking open snail shells rather than coconuts.

Chimps come a bit closer to humans in aiming and throwing things, as they have the same brachiator's shoulder joints that we do. They have been observed using both arms overhead to throw a stone down at the head of a dead monkey, trying to break open the skull. Also in aid of extracting a delicacy food (were brains the original *pièce de résistance*?). But there is a big difference in accuracy and range between braining a stationary target and exhibiting the skills of a Little League pitcher.

Actually, it isn't just late-afternoon-plus-fresh-air musings that generate all this fascination with throwing things. There is a good chance that the origins of language, no less, might be intimately tied up with the evolution of such throwing skills.

What? Is this an academic justification for the importance of baseball? Is he an athlete about to propose that the progenitor of talking man was a star pitcher? To ally any fears, let me assure you that I have impeccable credentials: I never graduated from the Little League, I haven't watched a baseball game in twenty years, I don't even know whether the Yankees are still in New York City. And someone else usually knocks the stones off the log before I do. Honest. But I suspect that pitching and talking did have something important in common, not recently but back a few million years ago.

It probably all started with the invention of one-armed rock throwing, handy for hunting prey without the usual long chase scene. Throwing possibly established the first important lateralization of a function to the left brain, an ability to rapidly orchestrate muscles in novel sequences. And I'll bet that this muscle-sequencing lateralization, most noticeable these days as handedness, was what started up not only toolmaking but language.

Where to begin? A boat drones by on a backdrop of sea haze. A seagull struts across the sand. Binoculars reveal a bald eagle perched in a distant tree, surveying its domain. The rabbits are out sampling the grass nearby.

Rabbits. Strange as it may seem for beginning a discussion of language (or, for that matter, throwing), consider a dog chasing a rabbit. One often sees a dog trying his best to head off a rabbit in a spirited zigzag chase. Since the domesticated dog does not have quite the motivation or the practice of his wild canine counterparts, the chase usually ends with the rabbit disappearing down a hole. Though sometimes, at least on this island, the rabbit is intercepted by an opportunistic eagle, to the dog's horror.

But the chase is the thing. And since it isn't taking place on a smooth racetrack, both hound and hare have a problem. There they are, running fast on all fours, but they cannot pick their way across the irregular terrain like barefooted bathers, deciding where to place each foot. No, they can only look ahead at where they'll be a second later and correct their more automatic four-legged gait away from the nearby tree or the imminent pothole. Essentially, they have to project their present position ahead in time, just as an astronaut computes the trajectory of a spacecraft.

And they do it well. Even when the rabbit tries veering sharply to one side, the dog often cuts the corner to head him off, instead of following the rabbit's path. Rather like the sheep dog's maneuver of trying to get in front of the straying animal. The dog is not only projecting himself ahead in time and space, but he is in effect "computing" the prey's trajectory as well—just as the astronaut computes the moon's orbital path and arranges things so that spacecraft and moon arrive at the same place at the same time.

But the dog and the astronaut both can make little last-minute corrections. If the dog doesn't turn as sharply as the sheep or the rabbit, he can perhaps correct this later. And the astronaut can engage in a little mid-course correction. Not as good as doing it right the first time, but fix-up-able.

Throwing rocks is a different matter—no mid-course corrections are possible, unless you've got one of those wire-guided anti-tank missles that an infantryman can steer toward the target after firing. But despite this shortcoming, throwing rocks has its distinct advantages over chasing a rabbit. It's faster and uses up less energy than a chase. But how did throwing rocks ever evolve from the usual prey-predator chase? If we are not to be Panglossian about things, what were the intermediate steps, and how did evolution come to reward them?

Actually, maybe throwing didn't evolve out of the chase. Or even out of hunting. Think back to that chimp, trying to brain that dead monkey. Suppose the monkey wasn't dead: braining it would make it safe for eating. One doesn't want to get too close to a wounded animal thrashing around—one might get hurt, and that's a no-no for evolutionary fitness.

So keeping a safe distance away and throwing stones at its head would have been a useful practice, keeping oneself healthy and one's offspring fed. Just move stone throwing up to the beginning, rather than waiting for dessert time. And the next logical step is to throw the rock in the first place, before the prey has already been brought down by other means.

But throwing a big rock overhead with both hands, while forceful, does not allow for much range. It would have to be a stupid or sick rabbit to allow someone to get close enough for a two-handed overhead throw. The way to get range is to throw a smaller rock with one hand. Uncocking the elbow, the way one does in throwing a dart, is the most important part of throwing: it's a sling, nature's version of the one David used on Goliath. But you've got to release the rock at just the right time.

(Want to know the physics behind why pitching is better? First of all, as the army discovered belatedly, bigger isn't always better. High velocity is better than massive weight—so the current M-16 army rifle now uses a small .22-caliber bullet but with lots of powder behind it. It's the kinetic energy packed by the projectile that does the damage—and while it is proportional to weight, it's also proportional to the square of the velocity. Thus the trick for throwing is to get a lot of velocity behind a smaller stone. There's a ninefold improvement when you triple the speed. But never mind—surely the hominid didn't understand the physics either, but still invented pitching.)

The hominid that started pitching stones would have a considerable advantage. It could now prey on animals that were faster than it could run. And what animal runs faster than a good fastball? The hominid could also avoid mixing it up with a prey, with all the attendant dangers of getting killed or maimed in the process. Lions do get gored occasionally by those wildebeest horns and elephant tusks. And throwing allowed larger prey to be taken on.

Action at a distance. One of the better inventions of biological evolution since sex. I don't mean to suggest that humans evolved out of an impatient chimp trying to eat brains for an appetizer rather than waiting for dessert time, but it does illustrate one possible path of inventions that could have led to throwing, and from that to a whole new set of possible prey in varied habitats.

But while this might be relevant to the origins of hominid hunting, whatever has it to do with language? Think back a few million years, to one of those walking-upright prehumans whose brains were one-third the size of ours. What set the stage for that rapid growth of the brain since the Pliocene? A rapid growth and remodeling which gave rise to a language specialization occupying a prominent chunk of the left hemisphere, and to parietal lobes which grew a lot beyond those of the chimpanzee. How did brain size lead to, or result from, such enhanced functions?

This may seem like a chicken-and-egg, which-came-first question. But such rapid change is, by itself, highly suggestive of the rapid exploitation of a new resource, or perhaps a new feature of the brain that led to a more-than-modest success in feeding oneself and one's offspring. Or perhaps brain evolution acquired a two-for-the-price-of-one mechanism, the way the two-pants-suit doubled the longevity of the suit jacket.

Maybe that new feature was not language per se but lateralization: avoiding the duplication of a function, just concentrating it in one hemisphere. While language skills might have been helpful for survival, maybe they were just a side effect of the lateralization of some other skill having more direct survival and reproduction benefits. Like maybe throwing.

Like maybe something else, too. Other than the nice supply of egg-size rocks on the beach, why pick throwing? Because sequential-movement skills, like language, seem lateralized to the left hemisphere. Language is "lateralized" to the left brain in the sense that for every fourteen people who suffer language impairment, thirteen will have left-brain damage and only one will have a right-brain problem. Such 13 : 1 odds are not matched by any other lateralized function. Visual-spatial skills are the specialization usually said to live in the areas of the right brain where language lives on the left. Damage to the right brain results in problems with getting dressed, or reassembling a flashlight when changing batteries, or reading a map. But spatial skills are impaired only two to five times as often by right-brain damage as by left-brain damage. So they are not lateralized as strongly as language.

Yet neither language nor spatial skills are notably lateralized in chimps; although some suggestions are beginning to appear, it is the ratio that matters. Something must have gotten such lateralizations started, something useful at the time. Something exposed to natural selection pressures.

So what is this about sequential-movement skills also being lateralized to the left brain? It was discovered that patients with left-hemisphere strokes had trouble fitting keys into locks, unlatching chains, and then turning doorknobs. With either hand. They could do each action separately, but doing them in the proper sequence was a problem. Right-hemisphere strokes don't usually create such problems. Doreen Kimura, a Canadian neuropsychologist, found it fascinating that such sequential-movement skills went along with the language skills of the left hemisphere—patients who were aphasic also might have difficulty using either hand for such tasks.

And then one of her doctoral students, Catherine Mateer, showed that aphasic patients also had analogous difficulties with facial movements. Again, the face was not paralyzed. Shown a color slide of Katy sticking out her tongue, the patient could mimic the action. But shown a three-part slide of Katy first sticking out her tongue, then pursuing her lips, and finally grinning, the patient couldn't act out the sequence in the correct order without making frequent mistakes.

So the left hemisphere seems to have a specialization for muscle sequencing, not only for hand movements but for oral-facial movements as well. Very interesting. Might that have something to do with language? Just because both skills are in the left hemisphere doesn't mean they are intimately related.

But in fact they are: The core of the human left language cortex is an area specializing in oral-facial sequencing. Catherine Mateer went on to Seattle to work with my colleague George Ojemann, using electrical stimulation of the surface of the human language cortex to test patients during epilepsy operations under local anesthesia. They discovered a wide area, just above and below the Sylvian fissure, where stimulation temporarily disrupts the patient's ability to mimic the facial sequences. And the very same area also serves to help the patient recognize the individual sounds that make up speech, the phonemes. Surrounding this area is a patchwork of specialized regions for semantics, syntax, and short-term memory for words. But that is another story (and is recounted in Chapter 16).

Maybe the sequencing lateralization is why language and right-handedness usually occupy the same left hemisphere. After all, what is handedness? It isn't fine movement control that distinguishes the good hand from the other. It is the skill in ballistic movements such as hammering. And throwing.

If movement sequencing is, literally, the core around which the rest of language is built, it raises a very interesting question, as Kimura, Mateer, and Ojemann have recognized: Which came first to the left brain, sequencing or language? Maybe the first lateralization wasn't language but rapid movement sequencing . . . ?

Back to throwing rocks at a log on the beach of San Juan Island. Knowing that my overhand throw is far more accurate than that of other apes, and that my muscles are being orchestrated by a sequencing specialization which resides in my left brain, I begin to wonder if the progenitor of talking man really was a right-handed star pitcher. Maybe it was throwing rocks at prey that got the brain started on the road to lateralization and language. So maybe that's why throwing is so important to human evolution: advanced pitching is distinctly a one-handed operation.

Unlike locomotion or the chimp's two-handed over-the-head throw, pitching requires using one side of the brain far more than the other during windup and throw. Any chance tendency for one hemisphere to be better than the other at rapidly orchestrating muscle sequences could make the animal better at throwing with the opposite hand. So pitching could have resulted in a strong selective pressure to permanently establish any lateralization tendencies for muscle sequencing that the genome might have randomly tried out (and, as noted in Chapter 1, maternal throwing success might have made that the left side of the brain more often than the right side).

But why this emphasis on throwing? Surely toolmaking would have rewarded lateralization, surely social cohesion would have rewarded language, surely there are many factors? All true, but the issue here is what got lateralization started, not its subsequent elaboration. Throwing has the advantage that it is one-sided, that it stresses the system for speed in a way that toolmaking doesn't, and that it has a really major payoff—an immediate one, promoting the survival of your own offspring, not just some advantages later.

Missed again. I get up and try stacking three piles of flat rocks atop the log, then retreat back to my comfortable seat, picking up missiles along the way. Ten straight misses, then a hit. Funny how you can tell a good one even before the rock leaves your hand. Of course, after it leaves your hand, it's too late to change anything—you have to make all of the corrections during the throw, not continually during the chase as the dog can do. So my brain has to compute, in effect, the trajectory of the rock and arrange the arc through the air so that it will come down on the target. Not bad.

But not too different (except for being compressed into a brief moment) from what the running hound and hare are doing, projecting themselves ahead in time and space, arranging it so that their feet do not insert into an approaching gopher hole. But there's nothing predominantly one-sided in locomotion, nothing to "reward" lateralization the way that throwing could have done by providing more survivors who carried its genes. Nor is projecting the prey's trajectory, when cutting corners during a chase, inherently one-sided. But maybe during throwing. . . .

Suppose I was throwing the rock at a moving target, as in one of those amusement arcades that dispense symbolic stuffed rabbits. Our ancestors weren't always throwing rocks at dead monkey skulls. Indeed, they probably developed a taste for rabbit and similar swift small mammals. Rabbits are used to being chased. They worry about rapidly moving animals on the ground. They are also used to worrying about birds swooping down out of the sky. But they aren't used to having rocks thrown at them. Or at least the rabbits and hamsters of a few million years ago weren't, back before man became the first action-at-a-distance predator.

Just move very slowly until you get into throwing range, as close as your experience tells you the animal will tolerate. Then cock your elbow behind you, so that your body blocks the animal's view of the first half of your throwing motion. When he does startle from seeing the rapid throwing movement, it may be too late—it only takes about a half second for the stone to reach most targets, and he can't move very far in that time. The object is merely to stun the animal, to foul up his escape long enough for you to grab him.

And the faster you can throw, the better. Not just to decrease the chance of the animal moving in the meantime. Faster is lots better. It means you can throw farther. And it means more stopping power. So that you can stun bigger animals, maybe bush pigs rather than just rabbits.

So faster is better. But much more difficult for the poor brain. The launch window for your missile is already narrow. To hit a rabbit from a car length away, you need to let go of the rock within a time equal to the time the shutter of a camera stays open when set at 1/200 second. And if the rabbit is twice as far away, your launch window shrinks eightfold. There aren't too many cameras around that operate at 1/1600 second. But your brain can time things that precisely, letting the rock slip at just the right fraction of a millisecond, over a throwing time of many hundreds of milliseconds. Somehow. It has to, if you can hit such a target. So saith Newtonian physics.

That's the sort of problem that motivates computer scientists to build bigger and better computers. But the human brain was getting bigger: if any random tendency in that direction resulted in better hunting, natural selection might have conserved those larger-brain genes, giving them a chance to try again.

Two targets down, one to go. At least children wasting an afternoon, throwing rocks at a log, would never bother trying to justify it. But now this throwing theory has a life of its own. It even makes the opposable thumb critically important—that's what lets the rock slip at just the right instant.

But while faster is better, is a bigger brain faster? A faster throw requires more precise timing, bringing in each muscle at just the right instant with an orchestra conductor's split-second timing. Neurons are notably noisy things and great precision is not their forte. But more are better. Heart cells in isolation will beat, but somewhat irregularly. Clump a few together, and they'll exchange currents so as to all beat together—and the beat will be more regular. Forty cells linked together are even more rhythmic, ticking along like a clock. The timing precision depends on the square root of the number of cells.

A similar problem arose when analyzing the circadian rhythm of the sand flea: how did they manage to restart their internal clock within ten minutes of exactly twenty-four hours? No individual neuron is that accurate, being more like sixty minutes off. But a simple circuit of many redundant neurons, all doing the same job, is much more accurate because it averages the output of the many neurons. Four times the number of neurons will reduce the fluctuations by half. Thus one simple way to make a neural circuit much more precise is to duplicate it over and over, dozens of times. Bigger is faster, and faster is better, as we noted before. Hence bigger brains are better?

Not bad. Now, if only I could just manage to reset the rhythms of all these local sand fleas, to make them think it was really the middle of the night rather than feeding time.

You know—oops, time to gather up more rocks—this might solve one problem which has plagued evolutionary theory. Sure, language is a good thing, but just how did it provide enough survival advantage to make all of those changes step-by-step? That's the trouble with adaptation stories. Fertile (or febrile) minds can always come up with a rationale for why something was useful or harmful. But was it exposed to selection pressures? Talking on the telephone can be a survival skill in rare instances—but riding motorcycles really exposes one to natural selection (negatively, judging from the severe head injuries seen in our emergency rooms). Granted, language might have helped a band of hunters coordinate the capture of a woolly mammoth, but what rewarded the gradual steps which preceded that stage of speech? Little improvements in survival skills, or an important jump into an unfilled ecological niche?

It is easy to see how such throwing improvements in hunting technique might have conferred an immediate selective advantage, perhaps of "new ecological niche" proportions. If our ancestors were anything like chimps, they lived in the tropics, gathering fruit and cracking nuts, grabbing some occasional meat—a way of life which doesn't work well outside the tropics. But suppose they invented a way of regularly, reliably eating small mammals and birds? Unlike fruit trees, that source of food is widespread, all over the earth, on the plains as well as in the woods, in colder climates as well as in the tropics. Hominids could have expanded their population to occupy new habitats. And then have gotten overextended, so that small groups were isolated by climatic changes and subjected to the environmental stresses which, in a small inbreeding population, may lead to a new species.

Predators-at-a-distance would be a new gimmick on the ecological scene, conferring a new niche. The traditional reward for the species that discovers a new niche is a population explosion. Just imagine the population explosion that will occur in the bacterial species that first learns how to digest nylon. First it will colonize the trash dumps and the used-car lots, then take over the shopping plazas. . . .

But there's also the two-pants-suit gimmick. Improved throwing skills would be a reward for the left-hemisphere skills—but surely the right half would be enlarged symmetrically at the same time. That enlargement would have gotten a free ride on the left hemisphere's accomplishment. The right-sided space corresponding to that occupied by the left-sided muscle sequencer would be sitting there, like an unfinished basement, just waiting to be put to good use.

That's two for the price of one. If the right brain found a visual-spatial use for the extra space, say improved depth perception, it might have again improved hunting—further reinforcing those bigger-brain gene combinations, and thus housing muscle sequencing in bigger quarters during the next round.

Another nice ratchet for evolution: Selective pressures that reinforce sequencing improvements in throwing also could provide extra space for visual-spatial, and vice versa. Thanks to lateralization, improvements go twice as fast? So perhaps the lateralization of throwing was what started the rapid enlargement of the brain. And perhaps "throwing genes" are little more than bigger-brain genes.

Undoubtedly humans are the outgrowth of a whole series of selective pressures which forced our predecessors to learn to eat new foods, live in new habitats, develop new reproductive strategies such as kinship. Each Ice Age has probably left behind a residue of skills which, like the boulders scattered across the landscape, remain after the vicissitudes have vanished. The generalized animal developed out of leftover specializations, just as the general-purpose computer developed out of such leftover special-purpose computers as the player piano and control systems for antiaircraft guns.

So, did generalized human language evolve from specialized hunting "computational" skills? Bootstrapping language through better throwing? If oral-facial sequencing built upon the throwing-sequencing machinery of the left brain, then it would be natural for the expanding repertoire of verbal expressions to also settle alongside in the left brain—and so to set the stage for more sophisticated language, hence culture, and science, and all the rest.

Is language, that *sine qua non* of humankind, merely a side effect of braining rabbits? Well, is that worse than its being an offshoot of the vocalizations used for threats and warning cries?

Those species-specific vocalizations in monkeys might, of course, also have served as the scaffolding upon which language was built. Certainly they are a more obvious choice than throwing.

But that seemingly logical proposition has one problem: the cortical specializations for those vocalizations in monkeys are not located anywhere near the Sylvian fissure, around which the human language area has come to reside. They are far away, near the midline of the brain, just in front of the motor area for the foot and just above the corpus callosum.

So perhaps language was built on the scaffolding of sequencing rather than snarls. From gestures to grammar. . . . Suppose that's why chimps are better at sign language than spoken language? Or why modern human males, who are often more natural pitchers, have language more strongly lateralized than modern females (though this is hardly to be envied: males are far more likely than females to suffer aphasia after a left-hemisphere stroke).

Finally! It felt right, and a second later, the rock knocked the last target off the top of the log. Well, not a bad afternoon's reverie.

So the key problem is a fast track for hominid evolution, whether throwing could have enlarged the brain faster than intelligence. Lateralization too is likely to lead to bigger brains, simply because of creating extra space opposite a successful specialization, the two-for-the-price-of-one ratchet.

Of all the known lateralizations, sequential muscle control seems the most central to the others, such as language. And what could have resulted in sequential muscle control residing primarily on one side of the brain? Well, an important muscle sequence involving primarily the opposite side of the body, rather than both sides equally or alternately. Say, handwriting or throwing or grooming or tool use. Surely handwriting wasn't the first. And no one has ever discovered a cortical specialization for grooming.

But what selective pressures conserved those useful gene combinations for sequencer lateralization and bigger brains, so that there would be even more in the next generation?

Throwing has big advantages over the other candidates. First, the "new niche effect": an immediate and fairly dramatic payoff in improved life-style, enabling new prey to be eaten with minimal risk, and the consequent ability of the species to expand into new habitats such as plains with lots of rabbits but no fruit trees.

Second is the "fastball effect": the muscle sequencer should have had a tendency to operate faster and faster. That's because higher-velocity projectiles would result in greater throwing range, better stopping power, and less time for the prey to react to the approaching rock. But once a short computation time becomes important—and more efficient hunting selects for it—you need some very fancy timing circuits. Maybe the brain reorganized itself to provide it, or maybe bigger brains in the next generation survived better.

The other candidates for an earliest lateralization—including language itself—have only slow, indirect, often delayed effects on survivability and habitat, nothing to compete with the exposure to selection pressures which throwing has via the new niche and fastball effects. And third, since throwing improvements could come from either left-sided sequencer or right-sided visual-spatial accomplishments, throwing would maximally utilize the two-for-the-price-of-one ratchet by driving it from either side.

If it all started with throwing, what were the following steps? Elaborating a hand-arm sequencer into an oral-facial sequencer could have come next. That should have been easy—borrow from thy neighbor. Remember the map of the body's muscles in motor strip, and how close the cortical neurons controlling face muscles are to those for the hand? They're next-door neighbors.

Then comes language building upon oral-facial sequencing. What selective pressures encouraged that, or did language just get a free ride for a while and settle alongside?—perhaps in some extra left-sided space, created by the most recent round of ratcheting having been a visual-spatial right-sided improvement in hunting. Surely the gradually increasing communications skills themselves provided a selective advantage for hand gestures and facial expressions, as well as speech.

Just as improved tool use with the right hand, say for chipping rocks into scrapers and knives, could have resulted from the rapid movement sequencer in the left brain. Again expanding the choice of prey via using sharpened spears, expanding the possible Ice Age habitats via tools to skin animals and sew. And, of course, the sparks that fly while chipping hard rocks were probably the origin of fire on demand. Good old throwing, useful for explaining everything.

But the capability for handwriting probably got the longest free ride off the throwing lateralization, until it got put to work a mere 5000 years ago. Whereupon the left brain and right hand created a written language, which allowed an accumulation of generation-jumping knowledge, and all that. Throwing strikes again.

And besides explaining minor things, such as how humans got started and what made book-learning culture possible, the throwing theory even reveals the true origins of baseball, establishes it as the most elegant of all sports, the fastball as the most fundamental of inventions. . . . Ouch! Sunburn. . . . Or is it sunstroke? Some daydream!

But it's all too much. No one will ever believe it. Except maybe baseball fans.

5

The Ratchets of Social Evolution

One needs to look near at hand in order to study men,
but to study man one must look from afar.

Jean Jacques Rousseau (1712–1778),
Essai sur l'origine des lingues.

Because the guides of human nature must be examined with a complicated arrangement of mirrors, they are a deceptive subject, always the philosopher's deadfall. The only way forward is to study human nature as part of the natural sciences, in an attempt to integrate the natural sciences with the social sciences and the humanities.

I can conceive of no ideological or formalistic shortcut. Neurobiology cannot be learned at the feet of a guru.

E. O. Wilson, On Human Nature

How is it that we can put ourselves in someone else's shoes, try to feel what he or she is feeling, experience things from another's perspective? Is empathy another one of those items on the human uniqueness list, something that we have but other animals lack? Or is empathy part of a continuous evolutionary development, a quality for which we can seek antecedents?

And so one must ask: What is the evolutionary advantage of being able to judge someone else's feelings? If you have ever lived in the Middle East and shopped each day in the *souk,* you will have seen such skills exercised and polished. There is nothing to prepare you for dealing with an American used-car salesperson like bargaining in the *souk;* you might even save the cost of a middle Eastern vacation if you practice enough. But upon your return, don't pick an expatriate Israeli, Turk, or Arab used-car salesperson with whom to haggle. For, like the shopkeepers in the *souk,* they are likely to be especially skillful at judging (from your eyes, from the expression on your face, from changes in your posture) whether hardening their price is likely to settle the deal or just send you off to try another dealer instead.

The oriental bazaar is just an organized version of the kind of bartering that has been going on for a long time, which is in turn built on top of the social give-and-take of primate societies. In the *souk,* the advantages to each party can be expressed in the artificial currency of shekels and dinars; in primate society, judging another's emotions is useful for establishing dominance hierarchies which, in turn, often determine access to food, mates, or the best nesting spot. A monkey that is skillful at judging when he can get away with something without a fight, or good at judging another's willingness to fight or acquiesce, is more likely to perpetuate his genes. Similarly, a mother who is good at judging an infant's needs from subtle clues is more likely to have an infant that survives to perpetuate her genes.

So it comes as no surprise that the human brain has a specialized region of cerebral cortex for recognizing the emotions expressed on another person's face. It had been suspected that the right side of the brain had something to do with this, as right-brain stroke victims sometimes lack an ability to tell if a family member is happy or sad. Now a group of researchers at the University of Washington in Seattle have shown exactly where it is. They studied epileptic patients undergoing brain surgery with local anesthetics. The awake patients looked at a movie screen in the operating room, on which the pictures of faces were projected. The faces were those of actors and actresses acting out a particular emotion: fear, disgust, happiness, anger, sadness, surprise, and neutral. The patient's job was to name the emotion expressed, simply matching the face to a list of the seven emotions which appeared on the following slide. A simple job, which patients do quite well over and over as the slides flash by. But on some slides, their right brain is electrically stimulated while they are viewing the emotional face.

In general, nothing happens. The electricity is faint enough that the patient cannot tell whether it is on or off. Only when one small area of the exposed right brain is stimulated does something happen: the patient makes a mistake, such as calling the happy face "disgust" or the fear face "anger." The area which seems to specialize in emotional faces is located above the right ear, in the rear part of the middle temporal gyrus; typically, it has nothing to do with their epilepsy and seems to be part of the normal brain. When the applied electricity confuses the neural circuits there, the emotional judgment is faulty—even though perception and memory tasks are not affected. And, when a stroke or tumor damages that piece of cerebral cortex, the defect in judgment for emotional faces may persist for some time.

It is not yet clear whether this is truly an "emotional" specialization or just an area for advanced spatial-pattern recognition tasks which detects other features besides emotion; indeed, the emotional faces specialization was not even what the researchers were primarily researching—the slides were merely being used to distract the patients during a memory task for tilted lines. But when they started analyzing the right-hemisphere maps that resulted, the emotional faces task was always disrupted at the same patch of middle temporal gyrus in one patient after another. There was no consistent bias to the errors—patients didn't tend to report happy faces as sad, or vice versa—but just judged wrong when stimulated.

Is this a right-brain speciality, or is there a symmetric representation in the left brain of this emotional faces area? That's hard to say, as researchers are always busy studying language whenever they get a chance to study a patient's left-brain functions. In the one patient whose left brain was studied with the emotional labels test, stimulation had no effect on such judgments (although it produced naming errors and affected verbal short-term memory when language tasks were tested at the same middle temporal gyrus sites). Certainly, the right brain is "dominant" for judging facial emotions. If we look at another person, it is the emotion expressed on the right side of their face (in our left visual field, which goes first to our right brain) which we register more easily than the emotion expressed on the other side of their face. Such experiments are done with so-called chimeric faces, where parts of two different pictures of the same person are pasted together so that one side is happy and the other sad. If the slide is flashed on the screen only briefly, we will tend to see it as uniformly happy or uniformly sad, depending upon which one was on our left side.

Are there such specialized neural circuits in monkeys, or cats, or frogs? In particular, just how much of this neural specialization do we share with the great apes? All but 1 percent of our nonrepeated DNA sequences are identical to those of chimpanzees and gorillas, suggesting that we may have inherited a great deal in common from our common ancestor. Accounts of chimpanzee behavior in the wild suggest that facial emotions are used extensively for communication; indeed, misreading an angry expression for a happy one would be fraught with implications for fitness. So it would not be surprising to find a similar area in the brains of the great apes—but no one knows yet. This is one aspect of neurobiology (another being language) where the human research is ahead of the animal research.

Judging another's emotions is quite important if you are thinking of cooperating. The major approach to the cooperation problem comes from game theory—not its gambling applications, but the attempts to use it to account for social behaviors. At first glance, evolution seems to be all about perpetuating your genes through your own survival and that of your descendants. So how can we account for self-sacrifice, as when someone saves another person's life at the cost of his or her own? Or monastic traditions (altruism is not just human—consider the bees that spend all their lives supporting their queen, who is the only one to reproduce). Most explanations for this (e.g., kin selection theories) note the genes in common between queen and drone, the tendency for self-sacrifice to happen mostly for close relatives—one's genes live on through their survival. How else could the denial of personal advantages evolve? But a broader way of looking at this is by asking how cooperation evolved, examining cooperation between neighbors as well as altruism between relatives, all of which suggests an important role for neural specializations which observe another person's face.

Cooperation in this sense occurs when you forego immediate advantages—where, for example, you yield the opportunity for food to allow another to eat. There can be long-term advantages to such cooperation: reducing the fights over food, with their spoilage of some of the edibles, the energy expended in threatening one another, the time wasted in maneuvering which could be used in searching for other food. But it works only if the other animal reciprocates next time. There are many advanced cooperative ventures (such as those discussed earlier as an alternative to chimp nut cracking in treetops); here, however, we are concerned with simple alternation and sharing, the first rung on the cooperation ladder.

The problem is often posed in simplified form as the Prisoner's Dilemma: two prisoners sharing a cell who can either fight each day over the single plate of food placed in their cell, or who can alternate or share the food. If they both cooperate from the beginning, fine. But if either defaults by taking advantage of the other, then both lose because of the continuing fights. And they are likely to get locked into noncooperation, eternally suspicious of each other. No question that cooperation is better—but it often appears to be unstable, not likely to survive. This simplified problem in socioeconomic theory has been extensively studied by Anatol Rapoport at the University of Vienna. Indeed, over 2000 papers have been written over the years by researchers on this dilemma and its philosophical implications!

Now if you are smart enough to figure out in advance that cooperation is better, and strike a deal with the other person, then there is no problem—but how did cooperation ever evolve in simpler animals if each animal is only looking out for itself (or secondarily for known relatives)? How could genes for cooperation gain a foothold if the rest of the population, the competitors for food, didn't have them and always "looked out for Number One"? In a world of individual success and failure, how were the seeds of cooperation nurtured?

One answer to this depends upon being able to recognize another animal as an individual—and being able to remember if he took advantage of your cooperation during your last encounter with him. If he did, you don't cooperate with him this time. Then you forget about it, not continuing to hold a grudge. This, you may remember from childhood, is called "tit for tat," and it is a simple yet powerful strategy which appears to have a good chance of evolving. Provided, of course, that the rewards for cooperation sufficiently exceed the usual yield from mutual noncooperation. And provided that there is a good chance of encountering the other individual again. Rapoport has applied it to the Prisoner's Dilemma, where there is an excellent chance of the two individuals having to deal with each other again in the future (in contrast, for example, to the chance of two strangers meeting on the sidewalk of a big city ever encountering each other again).

Provided that you cooperate on the first move, it solves the Prisoner's Dilemma: both prisoners will win through cooperation and will be unlikely to get locked into mutual noncooperation. But that doesn't necessarily say anything about evolution. Could "tit for tat" have evolved in a large group of interacting individuals? Is it stable once started, or could it be wiped out by enough cheaters? Robert Axelrod and William D. Hamilton have recently shown that Rapoport's "tit for tat" strategy sheds a great deal of light upon how cooperation could have evolved by Darwinian selection.

First of all, how could it have gotten started? It probably would have required a small group of individuals, likely to encounter one another again, some of whom had the "cooperativity gene." An isolated tribe with many relatives (and probably a lot of inbreeding so that even a recessive cooperativity gene would sometimes be expressed) would have been an ideal setup, but it really doesn't demand kinship in the manner of most altruism theories, just an isolated cluster of interacting individuals. It surely required a plentiful setting, such as the tropics, where making a living was easy, so that those who waited didn't starve. Later, when life inevitably got harder, the cooperation established in the bountiful years might have helped make the species more efficient in using scarce resources.

Could this trait have survived if the "tit for tat" individuals then joined with a larger group lacking the trait—the (probably apocryphal) trusting farm youth moving to the big city? Yes, say Axelrod and Hamilton, because the cooperating individual would lose only once to a me-firster before switching to noncooperation—but the advantages of cooperation would be manifest whenever he did encounter another cooperator. This lack of backsliding means, as they note, that "the gear wheels of social evolution have a ratchet."

The main requirement for the "tit for tat" strategy to survive is that an individual must not be able to get away with violations without others being able to retaliate effectively. In general, this means that you must be able to identify the other individual with sufficient certainty. And be capable of remembering him long enough to retaliate on the next encounter, whenever that is. And have a sufficient probability of indeed encountering the other individual again.

Obviously, a brain good at recognizing faces would be a better brain to benefit from the virtues of cooperation. There are areas of the human right brain that are used (though not exclusively) for remembering faces. Facial recognition is disrupted by electrical stimulation of the right brain and by many right-brain strokes; this involves areas of the right temporal lobe involved in memory for spatial arrangements (but not the emotional faces specialization noted earlier). For example, at the same sites where short-term memory for faces is disrupted (typically the superior temporal gyrus and around the back end of the right Sylvian fissure into the parietal lobe a short distance), memory for the angle at which a line was tilted is also disrupted. If the patient has trouble remembering the line's angle of tilt, he or she will also likely have trouble remembering a picture of a face after an eight-second distraction task. So improved memory for spatial arrangements could be quite helpful to a cooperation strategy since it would aid facial recognition. Or vice versa: A cortical specialization shaped by the cooperation-linked success of remembering faces might also come in handy for remembering the leaning trees which mark the route home.

A good long-term memory is another prerequisite for cooperation success under "tit for tat." We tend to take memory for granted without inquiring into the evolutionary pressures that might shape memory capabilities. A simple type of memory of obvious utility involves food aversion: avoiding a food which made you sick once before. The Garcia phenomenon occurs when an animal (something as simple as a snail will do) is fed a novel food; then, several hours later, a drug is injected that makes the animal ill. The animal, presumably thinking that the novel food made it sick, will avoid that novel food upon the next presentation, even if it is weeks later. It takes only this one exposure to produce a long-lasting memory, and so food aversion is a favorite example of one-trial learning studied by psychologists. For omnivores like humans, always sticking new things in our mouth to try them out, such a memory mechanism for taste and smell would be under strong evolutionary selection.

Many food dislikes in humans can be related to such a fortuitous pairing of a food with an unrelated illness causing nausea, and such food aversions can last a lifetime (a corollary is: Don't eat your favorite food when you are catching the flu—you might spoil its taste forever!). But if a food is going to make you sick, it will usually do so within a few hours after you eat it (one-month delays, as for the infectious hepatitis mentioned in Chapter 7, are unusual). So you need to remember what you ate, no matter what, for as long as it takes to react to it. For the snail, this memory starts fading after several hours, unless permanently embedded by the delayed aversive reaction to the food.

It is also easy to see the utility of memories that last a year—the deer remembering its migration path between summer and winter pastures. Animals lacking the two-hour food memory capability, or the one-year migration path memory mechanism, might not survive very well. But what is the comparable utility of a memory that lasts a week? What strong evolutionary selection pressures might first shape such a capability? One candidate is the individual recognition needed for cooperation strategies (and dominance hierarchies) to work. One telltale sign of "tit for tat" involvement would be if facial memories for cheaters were much better than for other individuals—if, like the food taste which preceded the nausea, the cheater's face were burned into the memory circuits. At least until after the next encounter.

Elaborations on the basic "tit for tat" strategy can be even more beneficial: Suppose that you can estimate, from the emotions expressed on the other person's face, whether to initially cooperate with them or not? Such an ability, presumably aided by the right middle temporal gyrus emotional faces recognition area, might cut your losses before they started.

But other variants on "tit for tat" may present problems. For example, too extended a period of noncooperation following a violation could, just as in the Prisoner's Dilemma, lock the individuals into perpetual noncooperation (forgiveness has an evolutionary virtue, no less). Axelrod conducted a contest of strategies submitted by sixty-two game theorists from around the world to solve the cooperation problem for which the Prisoner's Dilemma is an elegant metaphor; a computer simulated 200 successive encounters using each strategy proposed—and kept score. Many of these strategies were much more elaborate than "tit for tat," utilizing techniques dear to theorists such as Markov processes and Bayesian inference.

Yet the childish "tit for tat" strategy emerged clearly triumphant: it was not only evolutionarily stable, but was the most robust strategy tested. Thus the search for improvements on "tit for tat" initially focuses upon improving the chances within its basic framework: improving individual recognition, improving long-term memory, improving first-encounter prediction of poor risk individuals, improving chances of future encounters (by, for example, clannishness—preferring a local community in which most interactions take place with known individuals), and keeping strategies flexible (something that human frontal lobes do: people with extensive damage there tend to get locked into one particular strategy when approaching a problem and might never break their noncooperation strategy to try cooperation again, just as in the original Prisoner's Dilemma).

If this all sounds like a possible foundation for how social life developed—well, it may be just that. Attaining such an understanding is the strong motivation behind those 2000 papers on the Prisoner's Dilemma and all those game theory papers on "hawks and doves." Now there are some ways both for looking at the human brain's specializations and for appreciating the "social ratchets" that furthered them. The next decade should be a very interesting one for the interaction between neurobiology and social theory.

As Anne Roe and George Gaylord Simpson said in their book *Behavior and Evolution:* "It should by now be obvious that there is, indeed, a general theory of behavior and that the theory is evolution, to just the same extent and in almost exactly the same ways that evolution is the general theory of morphology." What may shape a wing may also shape behavior.

NEUROPHYSIOLOGY

On June 8, 1866, the official Censor's Committee of St. Petersburg presented an indictment against the pioneering Russian neurophysiologist, Ivan M. Sechenov, for attempting to publish a small book, *Reflexes of the Brain*: "This materialistic theory reduces even the best of men to the level of a machine devoid of self-consciousness and free will, and acting automatically; it sweeps away good and evil, moral duty, the merit of good works and responsibility for bad works; it undermines the moral foundations of society and by so doing destroys the religious doctrine of life hereafter; it is opposed both to Christianity and to the Penal Code, and consequently leads to the corruption of morals."

When the book was under investigation by the courts, a friend asked Sechenov (who eventually became Pavlov's mentor) why he did not seek the counsel of an attorney—to which he answered, "Why should I need a lawyer? I shall take a frog with me to court and perform my experiments in front of the judge; then let the State's attorney refute me!"

adapted from Leonard A. Stevens, Explorers of the Brain

The most modest research worker at his bench, pushing a probe into a neuron to measure the electric response when a light is flashed, is enmeshed in a large and intertwined network of theories that he carries into his work from the whole field of science, all of the way from Ohm's law to Avogadro's number. He is not alone; he is sustained and held and in some sense imprisoned by the state of scientific theory in every branch. And what he finds is not a single fact either: It adds a thread to the network, ties a knot here and another there, and by these connections at once binds and enlarges the whole system.

Jacob Bronowski, The Identity of Man

6

The Computer as Metaphor in Neurobiology

"Either you sit there and just close off, or if you do become engaged in what is going on with other people, then you have lost the thread [of your concentration upon your writing]. You've turned off the computer, and it is not for that period of time making the connections that it ought to be making.

"I really started thinking of my mind mechanically. I almost heard a steady humming if it was working all right, but if it stopped for a couple of days, then it would take a while to get it back."

Joan Didion, quoted by Sara Davidson in Real Property

Someday in the distant future, a team of archeologists will dig up a computer. But, alas, no one will have remembered to leave behind the instruction manual and blueprints. At a minimum, the specialists will open the cabinet doors and give Latin names to the various cable bundles. They may even embed a computer in clear plastic and slice it up with a bandsaw for closer examination of cross sections.

But if the computer is in working condition, another archeologist will undoubtedly try unplugging various cables, just to see what stops working. Through indulging such childlike curiosity, he or she would gradually identify the "head office" for various functions, such as the locations essential for input, output, memory, even addition. The psychological archeologists would put the machine through its paces, demonstrating that it could play chess and do payrolls simultaneously. Such relatively crude knowledge would parallel our understanding of the brain only a few decades ago, which was based largely on anatomy, pathology, and descriptions of stroke victims.

But locating the head office for various functions, whether in a brain or a computer, does little to reveal how the elementary computing units carry out their electrical machinations. The archeologist might want to attach a hi-fi amplifier to various computer wires, to listen in on the internal communications. The nerve cells that comprise the brain also run on electricity, computing and signaling with small voltages. A few decades ago, neurophysiology entered its "wiretapping" era: we began listening in on the conversations of individual neurons, the brain cells that analyze the external and internal worlds, signal the muscles, store information, and somehow provide the underpinnings of intelligence.

The Impulse Arises

The most eye-catching feature of a neuron's electrical repertoire is a sudden up-and-down-again swing in the voltage across the cell's membrane. Named the "impulse" by pioneering wiretappers, it shoots up about 100 millivolts (still, a thousand times smaller than household voltages) and then right back down again. Hence its nickname, "the spike." The whole event lasts about 1/1000 of a second, less than the time it takes most camera shutters to open and close. Making a light brighter or a sound louder won't change the neuron's impulse: in any given cell, they either set off a standard-size impulse or they don't. Nothing in between. And there is a threshold for setting it off, rather like pulling the trigger on a gun. A gentle pull and nothing happens; just enough, and bang; harder, and one still gets the same old response.

This all-or-nothing attitude on the part of the brain cell has long reminded observers of the computer's elementary computing unit, the flip-flop: information isn't coded by grading the size of the voltage, but by whether it is on or off. And indeed the neural impulse does seem more like flipping an ordinary light switch on and off, rather than like adjusting a dimmer switch. But what serves to flick the switch? Each neuron receives small voltages from hundreds or thousands of other neurons; it generally requires a sum of many such inputs to set off an impulse. Those more familiar with computers than neurons immediately see an analogy (physicist's fallacy #1): The neuron must, so the extrapolation goes, be detecting simultaneous events as does an AND gate. So the brain is a digital computer in disguise!

But this digital view of the neuron's physiology is simply wrong. Many flights of fancy have, however, been based upon it. They may describe some hypothetical computer, but not the particular one inside our heads. Its building blocks work quite differently.

Neurons Without Impulses

It comes as a heresy even to many neurophysiologists, but the fact is that many neurons get along fine without impulses—such as 99 percent of the neurons in the eyes that are reading this essay. Like deposits and withdrawals from a savings account, it is the balance between excitatory and inhibitory inputs that counts. In a nonimpulsive neuron, the balance controls the signal sent to the next neuron in the chain—just as the savings account yields ("outputs," in computerese) interest proportional to its balance. Neurons even have idiosyncratic rules like banks, some requiring a minimum balance before starting to put out anything ("Everything over $1000 earns 7%!"). So too with the product of neurons, the neurotransmitter molecules that are sent sailing on to the next brain cell: their output rate is controlled by the voltage balance, provided that it is above a threshold. Yet there is one big difference: Nonimpulsive neurons always send their "interest" on to another cell rather than adding it to their own balance. No compound interest.

A nice feature of some neurons, which one wishes the banks would also learn to mimic, is to provide output with no input: they spontaneously leak neurotransmitter molecules, thus allowing them to adjust the leakage rate up or down in response to even the smallest input. But not all banks are the same—and certainly neurons are a varied lot.

Inhibition ("withdrawals") is generally produced by a different neurotransmitter molecule than excitation ("deposits"). But neurons cannot add and subtract different kinds of neurotransmitter chemical. That would be like adding apples and oranges.

The solution, discovered in evolution at least 500 million years before the invention of money, is to first convert everything into an artificial common currency, like dollars. The neurotransmitter merely alters the leakiness of the cell's membrane, which raises or lowers its voltage. Volts, in the case of the neuron, are the medium for the message. And volts is even a fairly universal currency (some regions of some cells happen to have some local currencies, such as internal calcium or hydrogen ion concentrations, but they accept volts as well).

The Long-Distance Impulse

So, if variable volts are so useful, why are standard-size impulses ever used? Impulses seem to be essential for sending messages over a long distance (to the cell, that's perhaps a typewriter space or more) because voltage "balances" would be gradually lost by the leaky membrane. Just as one cannot use too long a length of leaky garden hose and still expect water to flow reliably out the far end, so lengthy cells may gradually exsanguinate the signal. Just as some lengths of garden hose are too short for the leaks to matter, so one talks of "short neurons" and "long neurons."

But long neurons have invented a solution: They can locate booster stations every millimeter along their leaky "axon" between the input end of the cell and the output end, where the neurotransmitter is actually released. The boosters aren't exactly high-fidelity amplifiers of the sort installed in underwater telephone-cable booster stations, but they are good enough to boost an impulse back up to its standard height. Thus the impulse arrives at the other end of the neuron none the worse for its travels.

Now there is no limit to how long the cell can be, provided it keeps adding another booster station every millimeter (one neuron may extend from your toe up to your neck, using perhaps 2000 booster stations called "nodes of Ranvier"). Cells short enough to avoid exsanguination may not require impulses for long-distance communication, yet they may use an occasional impulse anyway. Just for emphasis, short neurons may produce an extra squirt atop the steady secretion of neurotransmitter released by nonimpulsive means.

Coding with Impulse Trains

There are people who see codes and ciphers embedded in the impulses. True. One impulse, by itself, is like one dot or dash in Morse code: usually meaningless, except in the context of its neighbors. Thus there is much more to neural signaling than just what sets off the first impulse; it is the rhythm of the impulsive neuron that carries the message. But even the textbooks are fond of halting the story after explaining the first impulse—which is rather like a music critic listening to the opening strains of Beethoven's Fifth and then discoursing only upon the first note of dit-dit-dit-dah.

In the neuron, the brightness of a light or the loudness of a sound is indeed likely to be coded by varying the rhythm at which impulses are produced. Physicist's fallacy #2: The timing of an impulse train must be analogous to the on-and-off codes we send to "serial line devices" such as computer terminals and teletype machines. Ergo, the exact pattern of impulses must be very important, just as in computers. But some caution is in order: It would only have added insult to injury for the music critic's response to dit-dit-dit-dah to be instead a claim that the symphony was about the letter "V" in Morse code. (Besides, Beethoven came first by half a century and it seems more likely that Samuel F. B. Morse selected dit-dit-dit-dah for "V" based on the Roman numeral for Beethoven's symphony). There is, fortunately, a simpler encoding and decoding scheme to be found in most impulsive neurons.

Neurons that use impulses seem to work on analog principles, similar to their nonimpulsive cousins. When the impulse reaches the far end of the cell, it releases a standard-size squirt of neurotransmitter. So how do the inputs' voltages encode the message to be sent by the impulsive neuron? They simply vary the rate at which it produces impulses, with higher voltages causing the neuron to "beat" faster (a principle called frequency modulation, more recently applied to FM radios) and thus release more squirts per second. This makes the total neurotransmitter released each second at the distant end vary up and down as does the original balance of input voltages. If this sounds familiar, it's because this scheme is just a roundabout way of doing what the nonimpulsive neuron did, without impulses as the middleman: vary transmitter output with voltage input.

A better, more homely analogy for impulsive neurons is the sewing machine's pedal (rather than the gun's trigger). Press lightly and nothing happens. A little harder and the machine starts stitching rhythmically, but slowly. Harder still and it speeds up, proportionally to the pedal pressure—just like neurons speeding up and slowing down their beat as the balance of excitatory and inhibitory inputs changes. And just as some sewing machine models oblige you to press harder before anything happens at all, so some neurons have high thresholds, others low. Always diversity.

But some neurons have additional tricks, such as occasionally giving two impulses for the price of one, even though the voltage balance hasn't changed. Imagine a sewing machine occasionally doing a double stitch in the midst of its regular rhythm, or a bank offering to briefly double your money. It would be almost as startling as your heart giving a quick double beat. But it is just another item in the neuron's repertoire, probably encoding for something which the next neuron in the chain can decode—but which neurophysiologists still can't.

The moral of this little story? The workings of the brain are, at the level of elementary computing mechanisms, surprisingly simple—much more like balancing your checkbook and earning interest than like any binary mechanism inside a computer (or a physicist's fallacy). But there are enough variants to satisfy any free-market economist, to fill any ecological niche, to endlessly delight neurophysiologists.

Malignant Metaphor, Rampant Reductionism

Yes, the brain is a computer—that underpowered metaphor will have to suffice until a more subtle technological analogy is invented—but surface similarities to existing digital computers can be deceptive. There is a simple-minded longing within biology and psychology for hard-science insights (Freudians might call it "physics envy") that reduce soft-but-sophisticated phenomena to "nothing but" hard physics and elegant mathematics. (Perhaps like reducing Beethoven's Fifth to nothing but a Fourier series of different tones?)

Physics and mathematics can be most elegant, but the real world does not always cooperate with our aesthetic enthusiasms. Indeed, Ptolemy's belief that only perfect circles could be the building blocks of the universe sidetracked astronomy for fifteen centuries. Finally Kepler abandoned epicycles and tried out less elegant ellipses to represent the orbits of the planets (and then even Galileo didn't believe him—but Newton did).

Hopefully the aesthetics of computers and the appeals of reductionism will not similarly sidetrack neurobiology. Neural reality is much richer, more diverse, with many levels between membranes and brains. A level above membranes lie cells, with their regional specializations of different kinds of membrane. Above cells lie circuits of different kinds of cells. There is even one known cortical "module" with stereotyped circuity (a cubic millimeter known as a hypercolumn) repeated hundreds of times within the visual cortex. Somewhere above that, somehow connected, lies the high level of the uniquely human language cortex, whose core is surrounded by a patchwork of specialized areas for semantics, syntax—even a short-term "holding" memory for words.

Being able to explain one level in terms of the level below is a good test of the depth of your understanding, but it isn't everything (breadth, anyone?). While one might be able to reduce the chemist's valence to the physicist's electron waves, who would want to go to all that quantum mechanical trouble every time? Just as valence has a life of its own for predicting chemical compounds, so each level of a nervous system must be understood in its own right, in context, in all its rich variety—not just reduced to "nothing but" the level below. Nor to merely an analogy with computer hardware of recent design.

Evolution has, after all, been tinkering around with thinking machines for a lot longer than computer science has. Brains are the most elegantly organized bundles of matter in the universe. Why be satisfied with substitutes when you can study the real thing? And, as Oscar Wilde said, "Truth is never pure, and rarely simple." But certainly more interesting than malignant metaphor and rampant reductionism.

7

Last Year in Jerusalem

Nothing happens just as one fears or hopes.

Theodor Herzl

(*a phrase particularly apt for Professors' sabbaticals*)

My career as an Israeli tour guide began only five weeks after my arrival in Jerusalem to be a visiting professor of neurobiology. Not that there was any lack of natives, but I had a car—albeit the only car in Israel with Nevada license plates.

I had bought the orange VW Rabbit from a refugee from Las Vegas. I met various immigrants from the Soviet Union and Rumania—but in Israel, it is Las Vegas that is considered exotic. He had used it to transport his household belongings across the United States, across the Atlantic Ocean via the hold of the *Queen Elizabeth 2*, had driven it across Europe to Venice, whereupon he took it by ferry steamer through the Greek islands to Haifa. Considering that the port of Haifa was the only way to bring a car into Israel—all the land borders were closed to private cars then, this being just before the Egyptian border opened—it was remarkable how many different foreign license plates one saw on Israeli roads. But a blue Nevada license plate was unique.

Neurobiologists were not. Especially for two weeks in March when we had a lovely meeting at the university's new campus on Mount Scopus, which overlooks Jerusalem to the west, the Judean desert, Jordan River valley, and Dead Sea as you turn to look east.

In addition to the dozens of Israeli neurobiologists, there were more than a dozen invited lecturers from abroad. I was the first to arrive, since I was staying for a year. I found that the planning for the meeting was causing some problems around the Russian compound. (I must digress. The neurobiologists at the Hebrew University were housed in an old Turkish hotel which had previously served as administrative headquarters during the British Mandate, and then as the medical school after the university lost its Mount Scopus campus to the Jordanians during the 1948 war, which left Jerusalem divided. And, after a new medical school was finally built out in the western suburbs, the neurobiologists were given part of the stone-walled premises whose plaster was pockmarked from sniper bullets, the border having been just down the street from 1948 until 1967. It now had a noticeable odor of kerosene—the fumes from heaters which coped modestly with the cold and damp of the Jerusalem winter.)

Preparations for the Mount Scopus meeting were interwoven with the research in progress. And there were problems. For example, the lobsters on which I was to do research were not available yet. For a backup, they had ordered leeches (there is some logic to this, as the nervous systems of arthropods and annelids are surprisingly similar, so much so that some people think a mistake was made in making two different phyla out of them). The supply of leeches ordered from a European supplier hadn't arrived either. Undaunted, Itzhak decided that we would undertake a leech-hunting expedition on the Ramat Hagolan, which is familiar to newspaper readers and television viewers in the English translation: "Golan Heights." The decisive battles of several recent wars have been fought there.

And so at five o'clock one winter morning, we drove through the dark streets of Jerusalem in the zoology department's battered old VW van, past the walls of the Old City, and then started down the road to Jericho. From Jerusalem down to Jericho is about like descending from the rim to the bottom of the Grand Canyon: in the false dawn, the mountains on the far side of the Dead Sea even reminded me of the Grand Canyon. Halfway down, you pass a little marker alongside the road which says "Sea Level" in several languages. And in thirty minutes we were in Jericho, where the smell of orange blossoms permeated the morning air. We stopped at a gas station to fill up, passing the many Arab trucks waiting for the customshouse at the Allenby Bridge to open, so they could drive to Amman, Jordan, with their produce from the West Bank.

We drove out of town to the north, past a strange-looking hill: Tel Jericho, where excavations have uncovered one of the earliest towns known to archeology, dating back 10,000 years. And then we drove up the Jordan River valley (the Dead Sea is 400 meters below sea level; we were climbing toward 210 meters below sea level at the top of the valley), though with the Jordan River hardly ever in sight. This is an enormously deep river valley, and you expect to see a big, rampaging river with a flow suitable for carving such a valley, say like the Colorado River and the Grand Canyon. The river is very minor, even timid. The Jordan is a rift valley, the northern extension of the East African rift valleys in Ethiopia, Kenya, and Tanzania, where early hominid remains have been found. Those early-man places with names like Hadar, Omo, Turkana, Olduvai, and Laetoli are physically connected to early-civilization places like Jericho by that long rift valley. It was formed by the edges of two tectonic plates grinding away at each other, with maybe four major earthquakes every century to shake things up.

To see the winter dawn through the mists of the Jordan valley is to experience it as it must have been some morning at the end of the last Ice Age, as our ancestors were beginning to cultivate the wild grasses, to occasionally settle down near good water supplies such as the spring at Jericho.

We stopped for a mid-morning snack on the southern shores of Yam Kinneret, the Sea of Galilee. Then we drove around the lake, through Tiberias. There we first encountered Roman baths and UN trucks, the latter supplying the UN peacekeeping forces separating the Israeli and Syrian armies on the Golan. Then we headed up the Hula valley, the northern agricultural valley which lies in a horseshoe of surrounding mountains. The border with Lebanon is at the crest of the western ridge line; over the years, lots of rockets have come over that ridge, landing on the settlement town of Kiryat

Shemona ("Village of The Eight," named after those killed in an earlier attack). The border on the eastern side, however, used to be down just 100 meters from the valley floor, giving every Syrian soldier an opportunity for target practice on farmers. And a good view too, from their antitank fortifications on the hillside.

Actually, you'll never see the pre-1967 borders marked on an official Israeli map. But you can still guess where they were: just look for two dirt roads about 1 kilometer apart, running parallel to one another on the map but never crossing. Those were the old border-patrol roads on each side of the border. And in the Hula, the trees tell the story too: tall stands of trees were densely planted along the border road after 1948, to shield the farmers from the view of the Syrian soldiers. When the Israelis talk about the need for defensible borders, they have that pre-1967 situation in mind, among others.

The Israeli army has fought its way up that hillside, however, first in 1967 and again in 1973 after the attacking Arab tanks overran the Golan. We drove up it, at a pace suitable to our ancient vehicle. And saw the enormous view open up of all of northern Israel, the Sea of Galilee, and southern Lebanon. Near the road, bombed-out blockhouses still abound. There are yellow mine field signs everywhere, warning of mine fields which have never been cleared. And this is where Hagana Tiva ("protection of nature," short for The Society for the Protection of Nature in Israel, their equivalent of the Sierra Club) conducts nature tours? And where we are going leech hunting? All in the interests of science.

Our first stop was next to the military base at the Kafr Naffakh crossroads. Someone, undoubtedly from army duty, remembered a creek that had leeches, running north of the camp. So we started hiking overland, following the creek. I started to cross a muddy patch—and sank in up to my knees. It was a tank path, indeed the whole area was a training ground for tanks to wallow in. Everyone else knew not to step in tank tracks, just as all farmers know not to step on other kinds of off-color mud. There were, fortunately, no sounds of approaching tanks (like fire engines, their lack of adequate mufflers is quite distinctive).

But no leeches were to be found under any of the rocks. We

continued following the creek, however, past some disused barbed wire which everyone else ignored. And soon I looked up from my leech hunting to discover that we were inside the army base, amid rows of parked tanks and armored personnel carriers. But no people. We had seen a lone guard at the front gate. We decided, after failing to find a single leech, that it was the better part of valor not to leave by the front gate. So we retraced our muddy steps the long way around back to our van. I was beginning to get the idea that there was another side to the visible security which so overwhelms the visitor, such as having your bags and packages inspected each time you enter a grocery store or bank: terrorists never attack army bases, only civilians.

We drove farther north—and encountered a cowboy on horseback, herding some cattle. Never in my travels through the American West (which is, of course, considerably east of Seattle) have I seen such a sight except at a rodeo. I had to travel to Ramat Ha-Golan.

Had the Israeli cowboy seen any leeches? Yes, indeed, there were lots in the cattle drinking troughs, back on the ranch, and he would lead us there. But he accepted an invitation to ride with us, turning his horse loose to find its way home. And the horse naturally beat us there. Dipping nets into the drinking trough was, however, unsuccessful. The leeches were undoubtedly in the mud on the bottom of the tank, which hadn't been drained and cleaned for a long time. But we were welcome to drain the tank. Which we did. Two hours later, we had six leeches—and the cowboy had a clean cattle trough. I thought the cowboy had gotten the better part of the deal.

On the road north once again, we stopped for several Druze men who were hitchhiking. The Druzes are Arabs with their own secret religion (who are often at odds with the other Arabs), and they are the primary indigenous inhabitants of the Golan. Itzhak had a long conversation with them in Arabic, then fished out the jar of captive leeches and pointed to them, when it became apparent he wasn't using the right common word for the species in that part of the world. Oh, yes, they knew where to find them. There was a pond, right near the road they wanted to travel on if they could get a ride. So in they climbed, gesturing directions to the driver as we bounced along.

We stopped alongside a big pond, complete with herd of cattle and a few goats. The Druze men led the way, and there were indeed many leeches. You just have to ask the right people.

The rest of Hagolan is also most interesting. It is mostly a high plateau with several peaks, Mount Hermon being the tallest. But the plateau slopes off to the east into the plains which eventually lead down the biblical road to Damascus. The plains are zigzagged with deep antitank trenches, scarring the earth everywhere, dramatic evidence of a level of warfare quite different from anything anticipated along the Jordan River valley. Defensible borders.

Back in the lab, I got an experimental setup going by playing repairman, as the electronics technician was off to the army again. Experiments like these are often done in big screen-wire cages atop a table, the screening meant to keep unwanted radio waves from being picked up by the neural recording wires. If the screen door wasn't properly closed, your nerve cell might seem to be emitting a political harangue—though, in fact, the attached amplifier was merely receiving a broadcast from Radio Damascus. Except for the radio station received, the problems were little different from neurobiology labs elsewhere in the world.

There was, however, a locked cabinet in the wall adjacent to my table, and it was not long before I discovered what it contained: the armory. Periodically, someone would come in, lay down an Uzi submachine gun on my table or chair, and unlock the cabinet to put the gun away. All the faculty and students shared in the guard duty for the Russian Compound, usually a few hours every month. I became accustomed to moving submachine guns out of the visitor's chair in a professor's office when I went to talk shop, or accompanying someone on patrol to continue a conversation.

My favorite spot in the Russian Compound, when I wanted to take a break from the leeches in the screen cage, was a park bench in the grassy area in the middle of the old hotel compound. One could bask in the winter sun while still protected from the breeze, and read a manuscript. Since the Israelis publish mostly in English, I was always being asked to read a manuscript and improve the English. Often I would walk a block down to the corner of the Old City, the walled portion of Jerusalem inside which everyone lived up until the last century. There was a bakery where I would practice my fractured Hebrew and then carry back a sack of salty bagels to the courtyard. There was a friendly dog, an experimental animal that had free run of the place, who liked the bagels too (some were indeed as hard as bones).

The Friday morning seminar was informal, often devoted to talks by graduate students on their current work-in-progress. They were somewhat nervous and my presence didn't help the matter at all, as the rule was that they had to speak in English if I was present. All of the students had learned English in school, and they read a lot of English in their work, but that is surely very different from giving a talk where you have to think on your feet in one language and speak in another. The faculty were happy to have me as an excuse, since they knew that the students would soon need to give talks at international meetings. But they would sometimes relent in the middle of a talk if questioning in English wasn't getting the question answered. There would be a long dialogue in Hebrew until the issue was resolved, and then back to English.

Part of the informality of the Friday mornings was that someone usually brought a cake. Friday was a half-day of work, as the shops would all close by about two in the afternoon as the last of the weekend shoppers ran around. Shabbat started at sunset, but the noisy diesel bus traffic would disappear well before then. Shabbat was quiet in Jerusalem. But if you had forgotten to buy bread, you could always walk over to the eastern section of the city where the stores were open. Saturday mornings, there were walking tours of the Old City and outlying archeological sites.

On Sunday nights there was another, more formal, seminar called the "Jerusalem Nervous Club," whose audience seemed to come from all over Israel and whose speakers might come from anywhere in the world. Quite often, someone would have driven up to Jerusalem from Beersheva or Tel Aviv or the Weizmann Insitute in Rehovot. I spoke there twice during my year; the occasions were memorable not only for the vigor and intelligence of the questions, but because after the second talk, I was presented with a poem—a funny critique of my topic, written during the talk on a paper towel by a professor who could obviously listen and do a few other things at the same time.

The planning for the Mount Scopus conference created some distraction from the leech research. There was a small problem. Both Passover and Easter happened to coincide with the dates for the conference, this fact somehow having escaped the planners but not the hotel reservations clerks. One of the earliest Hebrew slang phrases which I learned was "Vitamin P," approximately translated as "influence." To extract a dozen rooms in various pensions (and monasteries!) around town took a considerable amount of repaying favors and asking friends here and there to intercede.

So, the organizers' solution to the housing problem involved a room here and a room there. Which meant finding enough drivers to pick up the guests each morning and ferry them across town to Mount Scopus. Having a car, I was transformed from guest into host—and instant tour guide. And around not only Jerusalem but the whole country, as excursions were planned for the guests.

It all had really started several weeks before, with the arrival of the second of the invited lecturers—my wife. Several days before she was due to fly in, and I was to see if I could remember my way from Jerusalem back to the airport east of Tel Aviv, it happened. Six PLO terrorists came ashore on the coast between Tel Aviv and Haifa, killed a woman photographer in the nature preserve where they landed, then commandeered a bus on the coastal road (at that time Israel's closest approximation to a superhighway). Firing at cars as they sped south toward Tel Aviv, they killed a number of civilians. The Haifa symphony lost a beloved violinist. The army set up a roadblock at the intersection just north of Tel Aviv, known to all Israel as the place where the used-car lots are located. The bus was stopped during a long exchange of automatic-weapons fire; thirty-nine were dead inside and along the highway; there was a possibility (only later ruled out) that several terrorists had escaped. The first curfew in the nation's history was declared, covering the area north of the Yarkon River, and the army searched day and night for the terrorists while families stayed within guarded apartment buildings. There are only a few English-language newscasts during the day, and I carefully checked the clock to be sure to hear them.

When I started out from Jerusalem for the airport, the curfew had been largely lifted, though the traffic was moving slowly because of the frequent police checkpoints. They have portable spikes which they lay out across one lane of the road, forcing drivers who value their tires to slowly weave a zigzag course and come to a stop. Passports and driver's licenses were frequently inspected, destinations asked.

The airport had an armored personnel carrier at the gate, but otherwise it was business as usual—though that implies an unusual amount of invisible security. Katherine emerged, extra luggage and all, from the giant customs hall, and we made it back up to Jerusalem, getting lost twice and getting inspected only once. Our next experience with the airport was somewhat novel. Several months later we had to ransom a radio from customs. In order to put up a deposit for one year in dollars rather than in the rapidly inflating Israeli pound, we were sent from the freight docks to the passenger terminal, which was the only customs office that could accept dollars. With the correct slip of paper, we were admitted through the "no admittance" side door to the giant arrivals hall, completely empty in the lull between morning and afternoon arrivals of jumbo jetliners. While I left to make a quick round of the branch banks to acquire sufficient cash, Katherine remained near the customs office in the arrival hall. Suddenly there were alarm bells and a small army of plainclothes security men and women invaded the empty hall, leaping over luggage conveyors with their Uzi submachine guns and scattering throughout the great barn. Katherine would have been alarmed, except for noting that the clerk at the nearby travelers aid desk continued to file her fingernails, apparently having seen the drill before. Back in 1972, an impressive Israeli neurobiologist named Aharon Katzir-Katchalsky was killed in the same room, together with many others, when Japanese Red Army terrorists pulled weapons out of their luggage and attacked the tourists.

I had seen Katchalsky's brother several nights before. He is, or was then, the President of Israel, Ephraim Katzir. We had been working late in the labs at the Russian Compound. Hearing a band playing outside, we wandered out into the cool night air to see what was happening. Next door at the city hall, there was a school band. And there was a receiving line outside the door of the city hall with some familiar-looking people standing in it. Such as the mayor and the prime minister. We walked over, two Israelis and myself, to stand behind the receiving line, but were finally waved back by a soldier who was the only visible security person. So from 10 meters back, we watched President Katzir arrive. I could not help but reflect on a similar scene in the United States, where spectators would have not gotten within a city block of a small security army. And this informal security was on the old border where the anti-sniper wall had stood when the city was divided between 1948 and 1967, a part of the city next door to Arab Jerusalem, not in a Jewish suburb.

And then I went to a reception at the president's residence, Beit Hanassi, an annual event for visiting professors. I remembered to take my passport for identification and approached the guard at the gate, briefcase and invitation in hand. He didn't look at anything, just waved me inside impatiently. I had a pleasant chat with President Katzir about a brain research organization in which his brother had been a major figure; he had been following its activities with interest, having the advantage of having been a chemistry professor before becoming the chief of state. There was only one security man visible anywhere—a point which I will re-emphasize so that the "siege state" image that everyone gains from the popular press will have another side. Israel has many sides.

We neurobiologists at Mount Scopus had the best of meetings, full of the warmth of the people and the springtime sunshine. The only moment when security concerns really entered my mind was, however, suitably dramatic and personal. It occurred while I was giving one of my lectures. The door at the rear of the room opened and in crept a soldier with submachine gun. Mind you, I was quite used to seeing soldiers on the streets carrying weapons: soldiers on leave are required to keep their Uzi or M-16 or whatever with them, so the sight is rather innocuous. They aren't even on guard duty.

But the soldier entering the door at the rear, whom no one but me could see, was obviously on active duty—I had learned by that time how to tell, by the suspenders supporting the ammunition belt and canteen. I realized, in midsentence, that maybe I had a problem.

My professional training had not exactly prepared me for lectures being interrupted by armed soldiers, presumably looking for something hostile. Should I catch the eye of one of the Israeli professors, or. . . . But a split second later, I smiled and waved. Beneath the grime, the soldier in field gear was an old friend from Beersheva, due to lecture at the meeting the next day. Except that his reserve unit had been called up. Like Easter and Passover, army schedules take no account of Mount Scopus conferences. Mike's appearance in the doorway meant that he'd finally been able to persuade his commanding officer to give him some temporary time off, and that we'd get to hear his lecture the next morning.

The conference was on the topic "Interactions Between Neurons and Target Cells," a title which takes in a lot of ground. Most Israeli neurobiologists attended; counting the twelve foreigners, we perhaps numbered thirty-five. For a small country, Israel has outstanding science; neurobiology is one of their best areas of expertise. I know excellent neurobiologists in several of the surrounding countries, who struggle largely alone with no one with whom to talk, and for whom even getting the scientific journals through the mails is a struggle. Israel's neurobiology outranks, in many areas, that in all but the largest European countries. And it is largely because the professors really care about their students; they see that their superb students are thoroughly trained, they send them abroad for years of postdoctoral training, they work hard to create jobs for them to come home to in Israel. For any developing country to have more than just a few scientists trained elsewhere, it will have to build up in much the same way. To have the end result also be of such high quality—well, that probably requires an educational system of real depth, a society that values learning, an ability to attract learned immigrants, and some measure of good luck. Israel has those things.

Like their political traditions, the academic traditions in Israel owe a lot to Britain—so the speakers were often interrupted with penetrating questions and occasional expressions of disbelief. A livelier meeting I have seldom seen. Between sessions, we gathered out in the warm spring sunshine and sipped coffee. I tend to remember facts by associating them with the place where I learned them. I am forever recalling dismal hotel lobbies and conference rooms, but I am surprised how often I recall Mount Scopus in remembering all the science I learned there.

Then I played driver and tour guide to some of the other visitors. The highlight of touring was undoubtedly the weekend descent into the Jordan River valley, down the road from Jerusalem to Jericho and hence along the Dead Sea to Herod's plateau-top palace at Masada. Driving through the Arab areas is not a problem; the troubles come across the borders, and there are strips of plowed earth and mine fields along the Jordan River, though minor by comparison to the defenses at the border with Syria.

What surprises most visitors is the extent to which the Jordanian and Lebanese borders are open. Thousands of Arabs weekly cross the Jordan River, visiting relatives or attending colleges in various Middle Eastern countries. Trucks filled with West Bank produce cross over to the markets of Amman. Relatives of West Bank Arabs, citizens of various countries at war with Israel, are admitted for three-month stays and wander around Israel like the rest of the tourists. Checking the freight and the people for weapons is routine, but mistakes happen. A Russian-made rocket, fired from somewhere in the West Bank, landed a kilometer from my apartment building, digging a large hole in someone's backyard, breaking windows and scattering laundry hung out to dry. The city government promptly sent some men out the next day to plant a rose garden in the crater. But the border stays open, as Israelis doggedly try to make friends with their neighbors, try to demonstrate that they are not really the devil incarnate of the propaganda broadcasts.

It is walking around Jerusalem that dispels the tensions which the two-week tourist might feel. The tourist might tend to worry about those armed soldiers on the streets, not knowing that they are heading home on leave, having to lug their rifles along. The tourist might jump at a loud noise, thinking it a bomb, not realizing that the rocky ground forces the construction industry to use a lot of dynamite in town. Once, when I was absolutely certain a bomb had exploded nearby (because I saw people running in great excitement down the street), it turned out to be a blowout of an overworn bus tire. (The closest I came to a bomb was, in a way, after I returned to Seattle. Watching the morning news on television, I saw some film footage of a bomb being disarmed in Jerusalem. The scene looked familiar. I finally recognized it as along a shortcut path that I habitually took from the labs to pick up falafel for lunch.)

But walking around, one rubs shoulders with people who don't act worried, who stop strangers on the street to ask the correct time or to ask directions. In walking ten blocks from my apartment to the lab at the Russian Compound, I would often be stopped three times by people asking directions—sometimes by tourists, but often by Israelis. The standard way of getting from here to there is apparently to walk out to the street, stop the first person you see, and ask directions. Then you repeat the process at every street corner, gradually zeroing in on your destination. And the different types of people you encounter in Jerusalem—Europeans, North Africans, South Africans, Russian and Vietnamese immigrants, all manner of Middle Eastern types—is far more varied than on the streets of London or New York. People walk in parks at midnight. That is not the behavior of people worried about violence in the streets.

The best part came after the Mount Scopus meeting was over and the visitors were delivered back to the airport. Katherine and I took off for ten days in the Sinai, driving down the Dead Sea past Masada, past the saltworks at Sodom and the pillars of salt seen in the hypersaline shallows of the Dead Sea, down through the Negev to the Red Sea. The road to Eilat parallels the border with Jordan, which is out in the middle of the Arava valley, only a few warning signs and less than a kilometer away. There is a roadblock at the beginning of the highway, where a soldier asks you to wait until another car comes along, then the two cars drive down together and keep an eye on each other. The purpose of this, considering how lightly guarded the border seems to be, is probably to keep tourists from stopping and wandering out into a no-man's-land. If you're supposed to keep driving, and there's a companion behind or just ahead of you, you probably won't stop to get your picture taken standing on the border. This simple device no doubt saves a lot of trouble.

Eilat is the Israeli port city on the Red Sea, paralleled several kilometers across the valley by the Jordanian port city of Aqaba. They are at the head of a gulf off the main body of the Red Sea, variously called the Gulf of Eilat or the Gulf of Aqaba, depending upon whose maps you read. The west side is the Sinai peninsula, the east side is Jordan and then Saudi Arabia. Eilat has the aspect of a European spa combined with a busy port—scuba divers and bathers on one hand, oil tankers on the other. But, because of the expansion of the oil port, there are only several kilometers of Israeli shoreline left undeveloped (there are only about 10 kilometers total, between Jordan and Egypt).

Each day we drove south to a different region of the Sinai coast (this being while it was still occupied by Israel, before its 1982 return to Egypt), exploring the coral reefs with face masks and snorkels. It is hard to imagine just how colorful the underwater sights can be—all manner of varicolored fish, fans of coral, an octopus here and there. And above water, the incredible mountains on the Saudi Arabian side towering over the gulf, facing the rugged terrain of the Sinai behind our beaches. Back in Eilat, we stopped in at the university's marine biological research station, talked shop, and got advice on research animals for a friend. The Red Sea has a red and white mollusc called the "Spanish Dancer" whose elaborate swimming motion (for which it is named) is probably controlled by the same group of nerve cells that have been investigated in a related nudibranch mollusk in the San Juan Islands near Seattle.

We eventually drove all the way south, to the very tip of the Sinai peninsula at Ras Muhammad. Israel managed this bit of occupied Egyptian territory as a sort of national park, with a park ranger to advise you at the entrance gate. We drove up and were greeted, in English, with "Ah, from one desert to another!" It was only a few minutes later, as we were driving along the washboard road, that we realized that he had been referring to our Nevada license plates. I might compare the Sinai to Utah or Arizona, but not Nevada. And the only thing comparable to the Red Sea isn't located anywhere close to the American Southwest. The Great Barrier Reef north of Australia is the only place with comparable underwater scenery.

Standing at the Y-junction of the Red Sea with the Gulf of Suez and the Gulf of Eilat/Aqaba, one looks west across the Gulf of Suez to Africa where, beyond the range of coastal mountains in Egypt, lies the Nile. To the east is the Arabian peninsula. To the north is Mount Sinai. To the south, the Red Sea stretches toward the Indian Ocean and the horn of Africa.

And one sees a fascinating contrast in geology. The tectonic plates of Arabia and Africa are separating all along the Red Sea, which is why the view from space makes the horn of Africa look like a piece of jigsaw puzzle detached from the Arabian peninsula. The spreading, with new crust welling up from the depths of the earth, is probably much like that which is taking place in the bottom of the Atlantic Ocean and which separated Africa from South America. Spreading can give rise to a basin-and-range landscape much like that seen in Nevada and Utah, which are also spreading apart. Looking northwest, one sees the characteristic shallow western slope of the Sinai mountains, gradually descending into a wide sandy beach along the approach to the Suez Canal. But looking up the other gulf to the northeast, the mountains come abruptly down to the water on both the Sinai and Saudi Arabian coastlines. This is where the old East African tectonic fault line goes, whose northern extension runs up the Arava valley into the Dead Sea and Jordan River valley, eventually up through the Sea of Galilee to Lebanon and Turkey. All the human history of the Sinai is but a brief footnote on the time scale of the hominid evolution that took place along this geological spectacle farther south, down in Africa.

Driving back north through the Sinai, the car broke down in the middle of the desert. Naturally. Where else? Inspection revealed that the washboard road had caused the engine to rock back and forth enough to fray the wires leading from alternator to battery, indeed to break the connection. My trusty pocketknife solved the problem temporarily. Later I tried to replace the wire and connector properly, but the part wasn't available anywhere. When I left Israel, the temporary patch was still holding tight.

The Jewish New Year is a special time to be in Jerusalem. For the final day, Yom Kippur, there is literally no vehicular traffic in most of the city starting at sunset and lasting twenty-four hours. A nearly full moon rose later that night over the walls and towers of the Old City. Without the cars and trucks and wheezing buses, it looked like a dramatically lit stage set from a biblical movie.

But the people made it real. Near a synagogue, an entire city block would be filled with people talking. The next morning dawned a beautiful autumn day, and soon people were out everywhere walking in family groups, visiting relatives and friends in the Old City, hiking all over West Jerusalem. Few people used the sidewalks; they occupied the middle of the street. The atmosphere was one of a people reclaiming their city—and perhaps themselves.

Thanksgiving is a holiday not widely celebrated in Israel. But we had a Thanksgiving dinner in the Old City, though with chicken substituting for nonexistent turkey. We were a collection of Americans and Canadians, together with a few Sabras. One of the professors had an apartment in the renewed Jewish Quarter of the Old City, looking out over the Temple Mount, the Western Wall nearby, with crowds of worshipers passing by below the balcony. The local vegetables and breads from the typical Middle Eastern *shuk* (or in Arabic, *souk*) several alleys away substituted for the cornbread and pumpkin pie.

Katherine—fortunately, as it turned out—had real turkey in New York, having been delayed in the States after a trip to a scientific meeting. Several days later, I had another trip to the airport to pick her up. The rains of winter were starting. The hot desert winds of the summer, the *shariv* or *hamsin,* which had lasted several weeks without respite, seemed long ago. On the way home from the airport, we did get in a visit to my favorite biblical ruin, Gezer, which I had earlier discovered on my own, rather off the beaten path.

Gezer is on a minor hilltop in the foothills leading to Sha'ar HaGai ("the gates of the valley") as one drives from Tel Aviv toward the mountain valley up to Jerusalem. Gezer has one of those views, like the one Michener describes in *The Source* (which is based on Megiddo, perhaps 150 kilometers to the north), which surveys the critical travelers' routes through the area, looking out over a rich agricultural valley to the north (the same "Vale of Aijalon" where the moon supposedly stood still for Joshua). It is within gunshot of the pre-1967 border, which is perhaps why it was never prepared for tourists after the archeologists had excavated it. It appears on the big survey maps of Israel, but they show no roads leading there. On my first visit, I had driven many kilometers down dead-end forest roads, seeking a way in. Finally I drove into Kibbutz Gezer and asked, in my limited Hebrew, directions to Tel Gezer. Oh, they answered in Brooklyn-accented English, just drive through our plowed field over there and head up the hill on its far side.

Not only are there cowboys on the Golan Heights looking like refugees from Wyoming, but there are farmers in the plains sounding like refugees from Flatbush. I do recommend a jeep if it has rained recently. But Gezer is spectacular, both for its views and its archeology, which includes ten monolithic pillars at which ancient pagan goddesses were perhaps once worshiped.

The cold spell naturally coincided with the failure of the heat in our apartment building. It took five days to replace the boiler; the standard one-word excuse that repairmen offer for delays is *milium* ("reserve duty")—which is sometimes even true, judging by how often my academic colleagues disappeared. Sitting around in the living room wrapped in a sleeping bag gave way to invitations to tea from neighbors with electric heaters. In the midst of which came a forty-five-minute phone call from the States, urging me to return for a visitation from the government people who dispense money for our research. No, they couldn't schedule it later, I really needed to return home four weeks hence. Right in the middle of a research project that needed intensive work, I had to pull up stakes and head home. Without the short vacation we had planned in Europe.

We took up jogging in frustration, doing laps around the nearby park. Except that I found that I couldn't keep up with Katherine. And sometimes couldn't even climb the three flights of stairs to our apartment without rest. Such, I thought, are the physiological manifestations of psychological stress.

On the day before Christmas, we finally got to take a long-awaited walking tour—the one that covers the eastern side of the city, up the Mount of Olives, along the ridge to Mount Scopus, then back to the western part of Jerusalem near the Russian compound. Because it is so long, it isn't often given. Our favorite guide for the municipality's Saturday morning free walking tours, Shera, had campaigned with her superiors for some time to offer it. My fatigue vanished and I walked 7 kilometers, up and down the hills of the city, along the northern walls of the Old City to the goat and sheep market, then down the hill to the Garden of Gethsemane, up the Mount of Olives through Christian shrines and Jewish cemeteries. We told Shera that we were unexpectedly heading back to Seattle, leaving in several days, and talked awhile about the virtues of Jerusalem and Seattle. Many Israelis are from somewhere else, but Shera seemed to be the only one who was from Seattle. She now lives down on the coastal plains at Kfar Sava, but frequently commutes up to Jerusalem by bus to teach and guide tours. There are many stories like that in Israel.

Christmas Eve is like most others in Jerusalem, as the Christian population is small. But since they started televising the Manger Square crowds in Bethlehem, sending them around the world live via satellite every Christmas Eve, half of Jerusalem seems to go to see and be seen. Thousands of cars stream southward to Bethlehem, which is almost a suburb of Jerusalem these days. It creates an immense parking and traffic problem: while Bethlehem is geared up for tourists, they usually arrive by tour bus. There has thus developed an elaborate system of invitations issued by the church authorities, and the Israeli police treat them as parking permits, simply turning back cars without them. To get an invitation now takes Vitamin P.

I spent Christmas Day selling my car. Transferring ownership was a full day's work for, as it turned out, three people: myself, the British professor to whom I was selling it, and a friend who came along to translate. Christmas is, of course, an ordinary working day in most of Jerusalem. We started out at the customs office, waiting in a line that had become familiar to me, as I had had to make numerous visits to keep up the paperwork of owning a car without paying taxes equal to twice the value of the car in the States (which an Israeli citizen would have had to pay).

It was while waiting at customs that Pat told me of his trip down to Bethlehem the night before. It had all started with a large and very rambunctious dog which Pat had somehow acquired months earlier in a weak moment. Said dog tended to knock over things at home and in the lab, when it followed Pat there. Pat had tried to train it to stay in the seminar room outside the lab entrance, so as not to frighten the rats inside and threaten the experimental equipment. The dog had a stiffly wagging tail, however, and so it churned the papers which I had carefully laid out atop a coffee table while rehearsing a seminar. I have a feeling that Pat's wife may have had a similar experience with the ever-so-friendly dog.

Pat gave the dog away to an Arab boy. But the boy couldn't take the dog home on the bus, so Pat offered to drive him home, using his wife's car. Pat was indeed anxious to get rid of that dog.

The boy, it turned out, lived on the outskirts of Bethlehem. And Pat forgot about the Christmas Eve crowds. So at roadblock after roadblock on the way to Bethlehem, Pat tried to explain to the police that no, he didn't have an invitation, he was just trying to take this boy and his dog home. The police were trained by the British, and Pat is the very archetype of a London professor (he even writes an occasional mystery thriller), so he eventually made it through to the Arab village in the hills outside Bethlehem. The boy set out with the dog across the fields, and Pat turned around to head back to Jerusalem.

But the dog broke loose from its new owner and came bounding across the rocky fields, happily chasing Pat down the road, tongue hanging out. Boy retrieves dog. Dog escapes again. Unrequited love. Rather a different view of Christmas Eve in Bethlehem than the one televised every year in its sameness. The Keystone Cops comedy enlivened the long wait on the hard bench at the customs house.

After the car had been transferred from my passport to Pat's—that is the phrase, as they stamp your passport with a big form which causes the airport people to ask about the car when you leave the country—we drove down to the motor vehicles office for the next step in the process. This office, with which I also had become increasingly familiar, was bedlam. Dozens of Israelis crowded along a counter, behind which were several harried clerks struggling with paperwork and shouting people. We had lost our interpreter by this time and were on our own. The filing system seemed to be quite elaborate. As we waited, two deep in the crowd, for our papers to reappear from the depths of still another filing system, I kidded Pat that, after all, the Israeli bureaucracy was inherited from the British. To which he instantly replied, "Yes, but I think the British were trained by the Turks." And he is probably right—there likely was an automobile registry back in 1917 when the Ottoman Empire still ran this part of the world, and the British probably just took it over.

My knees were sagging all through this, my fatigue having returned. Pat told me that the couple who hosted the Thanksgiving dinner were both sick with something. All I needed was flu. For on top of all of the packing up and last-minute running around, a new urgency had been added. Another long transatlantic phone call brought the information that my mother was to be operated on for cancer in several days. I changed my plane ticket destination from Seattle to Phoenix but couldn't move up the day of departure.

On my last day, I had one final chore, scheduled long before—to give a lecture at the medical school in Tel Aviv. I borrowed back the car, now adorned with Israeli license plates, and drove for a pleasant hour down the twisting road from Jerusalem to the plains, detouring by the airport on the way to transact some business with customs, as insurance against snafus the next morning before the flight out. Again, a visit to an empty arrivals hall. Though no one put on a drill of antiterrorist tactics for me, I did learn the secret back passage out of the arrivals hall, as they had to escort me up to the departures lounge, that being the location of the only branch bank then open which could refund my customs deposit in dollars. Enriched, I drove on to the medical school and a round of shop talk. Fatigue vanished again, in spite of the siege of packing the night before, and I had a pleasant, though preoccupied, day.

But it all went well and by dusk I left to drive back up to Jerusalem one last time. I stopped to pick up a succession of hitchhiking soldiers—which always meant explaining the American-style seat belts in the car—and we had the usual problems of where to fit in the assorted rifles. One really does "go up to Jerusalem" as in the Hebrew phrase, ascending from the plains, past the monastery at Latrun with its vineyards, through Sha'ar HaGai and then up a winding road still littered with burned-out armored cars, left over from ambushed convoys during the 1948 war. Then one crests a hill and sees Jerusalem on the next hilltop, a city of light and of decent people.

About five the next morning, the taxi drove us down through Sha'ar HaGai again. There was an almost-full moon shining over the valley where Joshua had commanded it to stand still. Beyond the road, one could imagine the landscape looking much as it did in Joshua's day, about 620 B.C. The valley has not always looked so peaceful. It has been swept by one army after another since Joshua; for nineteen years of recent history, it was an armed border. I looked carefully and spotted the silhouette of Gezer to the south, standing guard silently.

We cleared airport security quickly. I had my usual cup of cappuccino at the airport coffee shop. We got on the plane to Rome and I collapsed. Usually I get in four hours of work at the start of one of those long flights, but this time I slept. At the Rome airport, I again had some good cappuccino, though everything else was testing flat to me. Then I slept from Rome to New York. We hauled our seven suitcases up to the customs man at New York, expecting to have to defend our purchases of the last year. To my astonishment, he just said welcome and waved us through. Katherine had business in New York, so I continued on to Phoenix by myself with most of the year's luggage. And slept all the way. I arrived twenty-four hours after leaving Jerusalem, but flying west through nine time zones made it seven in the evening rather than four in the morning. My father met me. The operation was over, patient asleep, all's well, come over in the morning. I went to sleep again, this time in a bed.

The next morning, I cornered the surgeon and went over the pathology report. Yes, it was a tumor but the prospects were excellent. Then I went up to see my mother, trying to convince her that the statistics gave her an almost normal life expectancy. It is very hard to explain statistics to someone who might reasonably be worried; scientists and people who bet on horses may think statistically sometimes but they too, when worried, seek certainty. Then a quick trip to the main post office to arrange overnight delivery of some last-minute grant proposal revisions to Seattle. Then, sagging, I went back and slept.

But this time I didn't get up so readily. Soon my father was having to divide his time between visiting the hospital and nursing me. I still remember the days when I would stare at the irregularities in the white ceiling, and then see my father come in the door with a cheerful smile.

My wife flew out to investigate and hauled me in to see a physician. His nurse looked at me from across the room and asked how long had I been jaundiced? Hepatitis. And I hadn't even thought of it, fatigue and flue being easy explanations. The jaundice was then hard to spot, but I soon turned more distinctly yellow.

Later I remembered the people in Jerusalem who had gotten sick just before I left. I inquired. About eight of the people at the Thanksgiving dinner had eventually come down with hepatitis after the one-month incubation period, all of them Americans or Canadians by origin. Those born in Israel didn't catch it, either being immune since childhood or not having liked the food. One visitor in 50 to the rural Middle East gets hepatitis, compared to 1 in 350 visitors to Mexico or West Africa, or 1 in 9000 to southern Europe. When friends inquire whether I caught hepatitis eating spicy dishes in oriental restaurants, I explain to them about the American-style Thanksgiving dinner. You just can't win for trying.

I made it to Seattle in time for the visitation, though I was still notably yellow as I gave my presentation and answered questions. We got most of the money. A friend from MIT told me about the time he had come down with hepatitis—that it had taken him a full year to get his strength back, that he had been unaccountably sleepy and fatigued much of that time, but that one does eventually recover. The day after telling me that, he won the downhill ski races at Sun Valley, rather emphasizing the point.

And it was indeed a slow year. I haven't won any ski races since then, but I've taken several white-water trips down the Colorado River through the Grand Canyon. The river running through the mountainous desert didn't remind me of the Jordan, but the canyon certainly brought back memories of the Jordan River valley. I still think of that winter morning, winding our way up the valley north of Jericho, whenever I want to imagine our ancestors starting to settle down to agriculture at the end of the last Ice Age. Such settlements were the foundations of civilization.

Postscript

I hear, from another visiting professor just returned from a year in Jerusalem, that my poor car broke down in the Sinai again, causing Pat to be two days overdue in returning and worrying everyone. This time the car had been beyond pocketknife repair methods. Alas. And to think that the car was properly trained in the Nevada deserts beforehand.

8

Computing Without Nerve Impulses

Small Is Beautiful
Title of a book by E. F. Schumacher,
subtitled Economics as if People Mattered.
(Neurophysiologists still think large is beautiful,
but they are finding that small is different.)

With the fantasy world populated by personable robots named "R2D2" and the like, it becomes easier to think of our brain cells as fancy machines (there is even one neuron, in a seagoing slug, that was named "R2" back in pre–Star Wars days). Building a robot that can perform sophisticated computations about its environment via pattern recognition is a problem more easily appreciated than its converse: figuring out how an evolution-designed computer works. At the heart of this problem is understanding the computational abilities of the individual cells that comprise the brain.

Our usual first line of defense is to make analogies with foods: the special molecules called neurotransmitters, so popular with neurochemists and nonscientists alike, are indeed "tasted" (if not consumed) by the neuron. Their variety seems endless (there are now many dozens of types, when there were only a handful just a decade ago), and they are often close relatives of drugs and dietary substances. But to understand computation by a neuron, the somewhat more foreign subject of electricity is unavoidable.

However important chemicals may be for energy and for communicating between neurons, it may be said that computation within a neuron runs on electricity. To set off an impulse, the inputs impinging upon the neuron shift the membrane voltage about 10 millivolts until the threshold is reached. Like pulling the trigger on a gun, nothing happens until a threshold is crossed and something happens. In the case of the neuron, the something is the impulse (*en famille,* we call it the "spike"). Because various inputs must combine their actions to produce enough voltage (via the actions of chemical neurotransmitters) to cross the threshold, the take-home message perceived by most non-neurophysiologists (remember the "physicist's fallacy"?) is that the neuron functions in the manner of an AND gate or coincidence detector. This is unfortunate, as only a few specialized neuron types may actually operate in that digital manner.

Nonspiking Neurons

Although some neurons are functioning entirely without producing impulses, (remember those 99 percent of the neurons in the eyes reading this sentence) impulses are always used for long-distance communication, such as between eye and brain. (The distances within the hair-thickness of the retina do not often seem to require the amplification aspect of the impulse.) Voltage is still the common currency of the "nonspiking" neurons; it just regulates the release of neurotransmitter directly, rather than using the spike as a middleman.

When the strength of an input doubles, the voltage increases. This releases neurotransmitter at a faster rate. Inhibitory inputs decrease the voltage, thereby reducing the release rate. Thus the neuron can release neurotransmitter at a rate determined by the sum of many positive and negative inputs, just as a savings account pays interest depending upon the balance of deposits and withdrawals. This is analog computation, not digital. So, is the difference between spiking and nonspiking neurons the difference between digital and analog computers? No, because most spiking neurons are also analog in their computation, just using spikes for long-distance transmission of the result. The spectrum of neural computation and transmission processes can perhaps best be appreciated by first examining the so-called nonspiking neurons.

To function without spikes, a neuron must be small—and thus not one of the neurobiologist's favorite neurons whose large size makes them an easier target for inserting probes to measure internal voltages. Elongation over several millimeters usually means that the neuron uses impulses. But the brains of humans, as well as of our favored research animals, are all filled with cell types that fulfill the small-size criterion; most have yet to be studied. One attempts to extract principles of operation from the study of those neurons that can be reliably studied and then extrapolate them to the many situations, such as human sensory pathways through brain stem and thalamus, where they cannot be studied directly. The mud puppy, a primitive vertebrate, has a retina with exceptionally large cells; since John Dowling and Frank Werblin showed its extensive use of nonspiking neurons in 1968, many other vertebrate and invertebrate examples of nonspiking neurons have been reported.

Lessons from the Lobster's Stomach

But a better place to study nonspiking cells turns out to be located near the stomach of the lobster: the stomatogastric ganglion has just thirty neurons, enabling the researcher to get to know them as individuals and give them names. With a population of a classroom rather than anonymous millions as in the retina, one has the ability to study the interaction between a pair of neurons, just as a classroom teacher soon knows which children are exchanging secret messages under the desk tops.

At least one of the thirty stomatogastric neurons is nonspiking, refusing to fire a spike even under extreme voltage changes. However, by observing the downstream neuron, one can easily see a response. Because the synapse (the site of the functional connection between two neurons) is inhibitory, a positive voltage change in the "presynaptic" neuron causes a negative voltage change in the "postsynaptic" neuron.

By doubling the voltage change in the nonspiking presynaptic neuron, one can almost double the voltage response in the postsynaptic neuron. But if one halves the original voltage in the nonspiking neuron, nothing may be seen postsynaptically. There is a threshold for neurotransmitter release, a voltage below which the release rate is undetectible. But once above the threshold, more voltage causes more neurotransmitter release, which causes a more negative postsynaptic voltage response. So this nonspiking neuron is analog, but with a threshold. From less direct evidence, it would seem that many retinal neurons have similar analog characteristics.

But Katherine Graubard, who extensively studied the lobster stomatogastric ganglion nonspiking neuron, demonstrated a far broader principle of nonspiking computation in collaboration with Daniel K. Hartline and Jonathan Raper. In experiments at the University of California San Diego and at the University of Washington in Seattle, it was found that even the spiking neurons of the ganglion were also using nonspiking neurotransmitter release. Because their threshold for releasing neurotransmitter was lower than their threshold for triggering spikes, fluctuations in net voltage could be communicated to other neurons without spikes—and when a spike was triggered, it added an additional squirt of neurotransmitter, just for emphasis. Altogether, quite a different picture of neuronal computation than the digital view of the "physicist's fallacy."

Computational Schemes

In addition to further depreciating an outworn early "principle" of neural functioning, the lobster experiments illustrate that the relevant dichotomy is not analog versus digital, nor spiking versus nonspiking neurons. It is spiking versus nonspiking computation. Some neurons, such as most cell types in the retina, use nonspiking computation exclusively. Others, such as the majority in the lobster stomatogastric ganglion, use a mixture of spiking and nonspiking methods. To broadly categorize cells as "spiking" or "nonspiking" would be to miss an essential point (an otherwise excellent symposium volume was entitled *Neurones Without Impulses,* something of a takeoff on the famous *Animals Without Backbones* though hardly expressing as fundamental a dichotomy).

It would be tempting to say that spikes are merely a method for handling the long-distance problem: that whenever the input synapses are more than a few millimeters from the presynaptic neurotransmitter-releasing regions of the cell, spikes are used to initiate an all-or-nothing event which can be reproduced at a series of booster stations along the way. While this seems to be true, its converse is not. Some small cells use spikes, even though voltage attenuation over the cell's short length would not seem to pose a problem. As in the lobster cells and some vertebrate retinal neurons, spikes may also be used for emphasis, not merely for long-distance amplification.

Is the Neuron Ambidextrous?

The neuron's fundamental role is to control its rate of neurotransmitter release, regulating it in accordance with its microenvironment. For pacemaker neurons, the regulation is largely hormonal: special messenger molecules delivered by the bloodstream to the vicinity of the neuron and then diffusing the rest of the way. For most neurons, the special molecule is delivered up close: it is released from another neuron just a membrane's thickness away from the cell membrane in a complex called the synapse. At such a range, they can hardly miss their target. Binding to special receptor molecules on the cell surface, they either open up ionic channels through the membrane, or secondarily regulate internal chemical reactions. Neither, however, allows input strength to be compared between excitatory and inhibitory inputs, which use different neurotransmitters and receptor molecules. To avoid the adding-apples-and-oranges problem, neurons use a common currency—measured in volts rather than dollars.

If a neurotransmitter-releasing region is nearby, its release rate of neurotransmitter molecules will be changed as a result. But if it is farther away, the voltage change will be smaller, the attenuation depending upon the membrane's leakiness and the cell's geometry (branching patterns can be quite important). Several millimeters is merely a useful outside limit, beyond which some other process (such as the spike) is needed to keep the electrical message from dying out.

Some neurons have transmitter-releasing regions only at a long distance from the input synapses; the motor neurons of the spinal cord, which activate the muscles, operate in this manner and thus require spikes (the cat's spinal motor neurons are the "classic" example of neuron functioning being the first to be extensively studied). Others have such distant release sites but also nearby ones, scattered among the same cell processes which receive the inputs from upstream neurons. The nonspiking neurons tend to have this nonsegregated arrangement which, if their transmitter release has a low voltage threshold, allows a single neuron in theory to perform many different computations, one at each of its release sites—there is nothing to make such nonspiking release uniform at each different release site.

This breaks down the long-prized functional unity of the neuron. In spinal motor neurons, for example, the anatomic unit is also the functional unit because the release sites are all segregated at the far end of the cell in the muscle, all releasing together when a spike arrives. But nonspiking computation allows for multiple computations to be performed in a single cell, as each site sees a different weighting of inputs. Certainly a cell can combine spiking and nonspiking computational schemes, and the lobster neurons often provide good examples. They also show that nonspiking computation may occur in an elongated cell, just not in the elongated portion itself. This relaxation of the small-cell criterion means that many more types of neurons in both vertebrate and invertebrate brains may using nonspiking computation in the manner seen in the small nonspiking neurons, and that such a neuron may be communicating different messages to different cells downstream from it.

Spikes in a New Light

While nonspiking computation is usually analog (though with a threshold), this does not mean that spiking computation is digital, as in the "physicist's fallacy." Indeed, it too is usually analog with a threshold. Most cells are like pacemakers, in that they can produce a rhythmic discharge of spikes at a certain rate. Changes in the excitatory and inhibitory inputs, however, modify the rate. For the release sites far away from the inputs, the only voltage change they ever see is the spike. The rate of transmitter release is thus controlled by the spike rate. Which means that the net input voltage varies the release rate just as in the nonspiking neurons—but by using the spike rate as a middleman.

Thus analog computation in spiking neurons is controlled by the mechanisms that vary the spike production rate when the net input voltage goes up or down. As noted in Chapter 6, this works much like the control pedal of a sewing machine. For gentle presses on the pedal, nothing happens. When its threshold is reached, the machine starts stitching at a certain minimum rate; harder presses, and it speeds up proportionally to pedal pressure. So too the neuron regulates its spike firing rate with net voltage from the inputs, having a threshold below which it does nothing, above which it grades firing rate with input. Spinal motor neurons work this way, a fact known since the 1930s. More recent studies of this neural oscillator by Daniel Kernell in Amsterdam, myself, and many others have established how neurons encode information for long-distance transmission.

"Emphasis" is sometimes superimposed upon the usual voltage-to-rate code in the form of "double spikes," which release much more than twice the usual amount of neurotransmitter at the distant release sites. This patterning of the impulse train (common in many normal neurons but exaggerated in epilepsy and chronic pain disorders) may be more uniquely a property of spiking neurons. But it took many decades of cellular neurophysiology before the patterning was appreciated—and the study of nonspiking computation is still only a few years old, even though Ted Bullock foresaw it in 1959.

Studies of neuronal computation schemes are probably only beginning. Their value lies not in how to build a faster robot or computer—designing from scratch would surely be faster—but in understanding what goes on inside a brain and how it came to be. Evolution builds neural machines in stages, trying variations on what already works well enough to survive the vicissitudes of selection pressures (brains are, as someone once noted, jury-rigged). Pittendrigh noted that adaptive organization is "a patchwork of makeshifts pieced together, as it were, from what was available when opportunity knocked, and accepted in the hindsight, not the foresight, of natural selection."

It looks as if both spikeless and spiking computations are basic stages which have survived well enough to form the foundations of the higher functions of the brain. They are both ancient solutions to the problem of adding apples and oranges.

9

Aplysia, the Hare of the Ocean

But the cortex is an enormous haystack and we are more likely to find our needles in some smaller bundle.

J. Z. Young, Programs of the Brain

This sea-slug [Aplysia] is about five inches long; and is of a dirty-yellowish colour, veined with purple. . . . It feeds on the delicate seaweeds. . . . This slug, when disturbed, emits a very fine purplish-red fluid, which stains the water for the space of a foot around. *Charles Darwin, Voyage of the Beagle*

It can look like an underwater bunny, munching happily upon some underwater greenery, with ears standing up waving in the water. But the ears are not for hearing. They are rhinofors, and they taste the water. And at closer look, this animal is more akin to an overgrown garden slug than to an undersized garden rabbit. It is *Aplysia,* the "sea hare." Sometimes described as a shell-less snail, it too is a gastropod mollusc, seagoing edition (pronounced, in case you're wondering, ah-plea-see-ah).

But unlike its shelled relatives, it will never wind up as a culinary delight substituting for the European snails known familiarly as escargots. It doesn't need a shell for protection: the secret of *Aplysia*'s success is that it tastes bad. Even sea anemones, which will eat anything, spit out *Aplysia.* In fact, its major known predator is the neurobiologist, who can be seen wading around in the intertidal waters off southern California, in the Gulf of Mexico, or in the Mediterranean searching for a prized specimen amongst the seaweed with the dedication of a truffle hound. Most of the *Aplysia* familiar to neurobiologists eat a different species of seaweed than the ones Darwin saw in St. Jago in 1832, and consequently have far more purple than yellowish color. You are what you eat.

If cows are sometimes uncharitably described as machines for turning grass into beefsteak, so *Aplysia* might be said to turn seaweed into knowledge about the brain. Its remarkable facility in this regard is hardly due to being especially smart, or athletic (stupid and sluglike are the descriptions that more often come to mind), or being adapted to a strange environment (it has not yet seen fit to crawl ashore in southern California). It became popular with neurobiologists because its nerve cells were so pretty. True, they are often pigmented a bright orange, hardly what one might expect for invertebrate gray matter (and in a purple-colored animal, at that). But the beauty of the *Aplysia*'s nerve cells lies more in their size than in their color: they are often ten to fifty times larger than the cells of mammalian brains. And neurobiologists, whose attempts to stick glass pipettes into them are sometimes reminiscent of spearing apples floating in a barrel, appreciate an easier target (neurobiology was revolutionized in 1936 when J. Z. Young discovered a nerve fiber in the squid more than 1 millimeter in diameter).

Yet much the same appreciation could be made of neurons in other snails, and indeed *Helix* and *Limax* have also earned an honored place on the neurobiologist's workbench (and, once the nerve cells have been removed and garlic added, upon a student's dinner table). What is so nice about *Aplysia* is the familiarity with which one can get to know its nerve cells as individuals. The cell named R2 is the biggest. There is only one per animal. It always lives on the right side of the abdominal ganglion. Its membrane beats electrically, producing a familiar "R2-like" sound (provided, of course, that one listens in by inserting a glass pipette into the cell and hooking it up to a hi-fi system). The shape of R2 isn't always the same from one animal to the next, but there is no mistaking that forked branching pattern or that wrinkled membrane. When the neurobiologist can get to know a cell so well, it becomes an "identified" cell and is added to the map of the *Aplysia* nervous system which adorns the office wall of many a neurobiologist.

Accustomed as we are to a society of unique individuals with their own names, this may not seem special—until you try "identifying" the nerve cells in a fancier brain. There are only two identified nerve cells in the fish brain, and none yet in higher animals. In *Aplysia,* there are almost a hundred identified cells—still only a small fraction of the total number of *Aplysia* nerve cells, but an especially useful subset. Some of them are sensory cells, detecting a touch or water current, conveying electrical impulses from the skin into a ganglion of many hundreds of nerve cells where decisions are made. Some of the identified neurons are the motor neurons in the ganglia which send electrical impulses out to a muscle, causing contraction and movement. But some neurons are simply decision-makers, interneurons with no branches coming or going from the skin and muscles. So sensory neurons deliver impulses to interneurons and motor neurons; interneurons affect the motor neurons and modify the messages sent by the sensory neuron endings upon the motor neurons. As more and more of these interneurons have become identified, neurobiologists have been able to figure out some of the basis for learning and memory in these animals.

One cannot expect an *Aplysia* to memorize a list of telephone numbers, so learning experiments tend to focus upon simple behaviors and how they come to be modified. If you touch an *Aplysia* anywhere near its gill, it withdraws and folds a flap of skin over the gill for protection. Touchy beast. But repeated attempts to touch its gill result in less vigorous withdrawals: the beast gets used to the stimulus, or as the technical term goes, it "habituates." Wouldn't it be nice, some neurobiologists mused in the late 1960s, to find the mechanism that altered within the nerve cells and caused this weakened response?

That started a long detective story. Maybe the sensory neuron was the culprit, giving a weaker response to the repeated stimuli? So a fine glass needle called a microelectrode was inserted inside a sensory neuron to measure the electrical responses. They didn't change with repeated stimuli.

Maybe the motor neuron or the muscle was giving less response to a standard stimulus from the sensory neuron? So the microelectrode was inserted into the motor neuron and an electrical current passed which mimicked the current produced by the initial sensory barrage. Every few minutes, the intracellular current mimicked a standard sensory barrage. The muscle contraction that moved the gill was examined to see if it was becoming weaker. Not notably, sorry.

Well, if it isn't the sensory neuron's barrage or the motor neuron's responsiveness that alters, what is left? The sensory neuron makes direct connection to the motor neuron, but it is possible to alter the strength of the connection. This connection site, called a synapse, uses chemical processes in a manner that the remainder of the cell avoids (most of the rest of the cell runs on electricity, pure and simple). When the electrical nerve impulse arrives from the sensory endings in the skin, it allows some calcium to enter the nerve terminals. That causes a chemical substance to be released (called a neurotransmitter; it is a simple molecule). That neurotransmitter spreads through the bodily fluids a very short distance to the motor neuron and opens up some channels through its membrane, which causes its voltage to increase. If that seems complicated, it is. Indeed, it offers all sorts of opportunities for nature to modify the system to produce learning. But, as it happens, only the first part of it is important for our story (if you yearn for more, start by reading Chapter 9 of *Inside the Brain*).

So next our neurobiological detectives tried examining the size of the barrage in the sensory neuron as observed from inside the motor neuron. And, while it hadn't changed as viewed from inside the sensory neuron, it was getting smaller and smaller when viewed from inside the motor neuron. The synapse was changing its strength. But how?

The easiest possibility is that less neurotransmitter was being released from the terminals of the sensory neuron. Because the space in between neurons is so small and hard to locate, this could not be tested directly—but a series of more indirect measurements suggested that there probably was indeed less release. So maybe it was just fatigue in the release mechanism, or maybe supplies of neurotransmitter ran short?

No, because the usual amount of release can be instantly restored by a simple maneuver: just deliver an electric shock to the head end or the tail end of the animal, far away from the gill, and the gill-withdrawal reflex will get much bigger. And how does that work? What neural pathway from head or tail manages to affect the terminals of the gill sensory neuron ending upon the gill motor neuron, so that they again release a lot of neurotransmitter? Fans of English murder mysteries may even recognize this involved plot.

To make a long story short, the culprit is serotonin or something related to it. Serotonin is a common neurotransmitter even in human brains where, among other things, it regulates sleep and wakefulness. And mood: serotonin imbalances are thought to be involved in severe depressions. Neural pathways from the *Aplysia* head or tail seem to release some serotonin somewhere near the synapse connecting the gill sensory neuron and the gill motor neuron, and that somehow serves to increase the neurotransmitter release and hence augment motor neuron responses and produce a more vigorous gill withdrawal.

The details of the augmentation pathway are even becoming obvious. Serotonin (1) affects the synthesis of cyclic AMP (2), another "messenger molecule" acting inside the neuron. Which in turn activates a protein kinase (3). It closes down a channel through the membrane used by potassium ions (4). That prolongs the next nerve impulse (5) in the terminal (potassium flows limit the impulse duration, so less is more), which allows more calcium to enter the terminal (6) and release more neurotransmitter (7). All clear? Even if it isn't, perhaps you will understand the power of the reductionist scientific approach which dissects the mechanism into its seven (so far) component pieces. That is how you come to understand how nature can modify systems such as these to produce learning and memory. Even the potential mechanisms which, it turned out, were not used in *Aplysia* learning (such as cyclic AMP effects directly upon calcium channels) may prove useful in analyzing other systems: once you learn how to eliminate the possibility in one animal, you often have the tools to check out other animals. It seems likely that some of the other potential mechanisms are used elsewhere. It will surprise no neurobiologist if those *Aplysia* results aid the study of human learning—and learning disorders.

But, you may say, surely habituation is a trivially simple kind of memory. Simpler even than Pavlov's dogs salivating to the bell alone, without food present. Can we perhaps get the *Aplysia* to respond to a Pavlovian paradigm, where we pair an innocuous stimulus with a more meaningful stimulus—and get an augmented response next time to the innocuous stimulus alone? Yes indeed, though it took a decade to learn how to do the experiment the right way.

Take two different sensory neurons leading from the gill to a gill motor neuron. Each produces a standard-size response in the quiet animal. Now shock the tail—but lightly, not so strongly that it greatly augments the gill-withdrawal reflex as before (or increases synaptic strengths generally, which is called "sensitization"). This is what prevented the following experiments from being done for a decade, until someone tried turning the shock strength way down. Like Pavlov pairing food with a bell, the researchers tried pairing the mild tail shock with a sensory barrage in one of the two sensory neurons. For us to say that associative learning took place, the tail shock would have to affect only that one sensory neuron whose activity coincided with the tail shock, not other sensory neurons generally.

And it worked. If the tail shock was given shortly *after* the sensory barrage, things changed for hours thereafter: succeeding sensory barrages in that one cell would produce much larger responses. That demonstrates a short-term memory of the impulse which is "developed" by the tail shock to cause a more long-lasting change. The synapse became far stronger. Indeed, even without further conditioning, the strength of the synaptic connection grew for many hours after the first pairing, suggesting that the results may provide a model for the consolidation of long-term memory as well as short-term memories in *Aplysia.*

And surprise: If the tail shock occurred *before* the sensory neuron's impulse arrived, nothing much happened. And they say, time has an arrow: Only sensory barrages preceding the noxious tail shock served to prime things so that subsequent gill withdrawals were augmented. There was something about the aftermath of the impulse in the terminal of only that one sensory cell which primed it to respond to the (presumably serotonin-mediated) message from the tail shock. The other sensory neuron, not itself activated before the tail shock but only at other random times, was modified only a small amount (i.e., there was thus a little nonspecific sensitization from the tail shock).

It is indeed a long way from such *Aplysia* learning to explaining associative conditioning in Pavlovian dogs, much less to memorizing telephone numbers. But at each stage of our analysis of human brains, we have typically found that simpler animals share the same phenomena. The nerve impulses that I record from human brains look just like those I record from lobsters (well, almost—I can tell them apart, but I'm an expert on irrelevant small details). They typically share the same underlying basic mechanisms: the same neurotransmitters, the same calcium channels, the same modular neural circuits (such as seen in the withdrawal reflex), and so on. And the same biophysical principles apply all up the scale.

Serotonin-augmented synapses in *Aplysia* may well be a model for one kind of human learning. It may someday even help explain some of the characteristic features of depressive mental illness (several of the leading neurobiological researchers in the invertebrate area are indeed trained psychiatrists). It is the research laboratories of one of those physiological psychiatrists, Eric R. Kandel, at Columbia University, which has been the focus of much of the *Aplysia* learning work just mentioned, with Tom Abrams, Bob Hawkins, and Tom Carew being particularly responsible for this latest breakthrough in paired conditioning. But there are many invertebrate researchers scattered around the U.S., Canada, and Europe (indeed, Ken Lukowiak's lab in Calgary and John Byrne's lab in Houston reported complementary results at the same time). It is this vigorously interacting community of workers that has made this field of neurobiology one of the most exciting areas of our research into how brains modify themselves.

But some credit must go to the "stupid slug" for having become such an accessible haystack.

10

Left Brain, Right Brain: Science or the New Phrenology?

If the Lord Almighty had consulted me before embarking on the Creation, I should have recommended something simpler.
Alfonso X, King of Castile (*1226–1284*)

Make it thy business to know thyself, which is the most difficult lesson in the world. *Miguel de Cervantes* (*1546–1616*)

Left, *right,* left, *right*—the marching song of the two-mind movement. To hear them talk, you'd think that everyone had a second mind, suppressed by the first. That the vocal left brain dominated the poor artistic right brain. Preventing it from getting a creative thought in edgewise. Soon there will be a consciousness-raising movement: Stop referring to the left cerebral hemisphere as the "dominant" one. Invent a more egalitarian term like co-chairperson. Co-chairhemisphere?

Alas. Were cerebral physiology so simple! If there were strong dominating influences, it would make our research far easier. It is unfortunate that "dominance" is a word with two entirely different meanings, even within psychology. When talking about pecking order, dominant refers to an animal that usually wins in a one-on-one encounter, the animal that can approach, threaten, and successfully displace another animal from food, mates, or the best nesting place. In talking about the cerebral hemispheres, however, dominant is merely a shortening of the technical term "language-dominant hemisphere." It is the outcome of a test to find out where language lives in a person's brain, such as injecting anesthetics into the left and right carotid arteries and seeing when the patient stops talking (or the simpler, but not as accurate, test that merely involves having the subject look at a dot in the middle of a screen and then briefly flashing words to the left or right of the fixation point; people with left-brain language will have an easier time with right-sided words since the information goes first to left brain).

Although a few percent of people have right brains that are language-dominant, about 93 percent of us use the left side. A few percent have "mixed dominance," where both sides are used for language (that is, injecting anesthetics on either side will interfere with speech). But the term hardly refers to language dominating art or music: it's just which side is more essential for language than the other.

Shades of gray become black and white when the dichotomizers go into action. But the real problem is that most of the creativity arguments have about as much to do with the brain as does the English language. The structure of the brain probably has a lot to do with the capability for, even the "deep structure" of, language—but brains hardly come in Chinese, Swahili, and English flavors. Like English per se, the creativity and holistic-thinking influences probably lie more in the realm of culture than brain structure. And hardly on a particular side of the brain.

Twenty years ago, similar suppressed-creativity arguments were floating around. It's just that they were then phrased in terms of contemplative Eastern thought versus authoritarian Western religious influences. More recently, the dichotomous rendition was holistic versus linear thinking. And now the mod metaphor is right brain versus left brain. Except that it is the worst of mixed metaphors, the kind that mixes up metaphor with reality.

Being a neurophysiologist, I suppose I ought to feel that progress has been made: in no other age could it have taken a mere twenty years to shift from a predominantly religious metaphor to a semi-scientific one. But the neurophysiologists and neuropsychologists who specialize in the human cerebral cortex are starting to view the left-righters with something of the wariness which the astronomers reserve for astrology.

In one sense, the picture of the mind painted by the left-righters is rather like one of those magazine illustrations of the human brain and its convoluted surface—one feels quite sure that the artist has never seen a real human brain, either a fresh one or a preserved one. The result of embroidering upon another artist's rendition bears even less fidelity to the original, an artistic version of the spread of a rumor. My favorite painting of the brain's convoluted surface is not the usual gray modeling-clay rendition but a red and purple rising-sun sphere whose "convolutions" come close to looking like the cracks in a parched mud flat. (It is my favorite only because it's on the cover of a paperback called *Inside the Brain* by Calvin and Ojemann; our publishers carefully avoided showing it to us until after the press run.)

If you try tracing some of the left-righters' enthusiasms back to the scientific evidence, you'll often wonder how the rumor even got started. It is not that the scientific evidence contradicts their notions, though that sometimes happens. But they've gotten way out in front of the state of the scientific art, in about the way the phrenologists' maps (all those political subdivisions on the brain map marked love, acquisitiveness, compassion, etc.) were premature, quite unsupported by any evidence at the time. Indeed, one has the suspicion that the nineteenth-century parlor-game aspect of phrenology resulted in pinning functions onto a brain map, like pinning the tail on the donkey while blindfolded. The left-righters' notions may have a base in a laudable know-thyself effort, but their standards for evidence can sometimes be subservient to their aesthetic enthusiasms ("a nice symmetrical arrangement, isn't this?"). To know thyself may be the most difficult of lessons; to know the brain is surely one of the most challenging of the sciences, one that does not lend itself to easy labels.

Besides the usual human tendency to see things as either this or that, the dichotomizers' view of cerebral physiology tends to be biased by the special case of language. Language in the human brain is almost totally located in one cerebral hemisphere. For every fourteen people suffering language difficulties after a stroke near the Sylvian fissure, only one will have suffered right-hemisphere damage. This 13 to 1 lateralization ratio for language is the strongest deviation from the usual vertebrate brain plan of about 1 to 1, namely doing most things pretty equally in the two halves of the brain.

But nothing else has such odds. Take visual and spatial relationships, the function most often assigned to the right-brain convolutions opposite those occupied by language in left brain. While there is indeed a specialized area used for recognizing whether a friend's face is happy or sad located just above the right ear, most visual-spatial functions are not exclusive to the right brain. A common one impaired by strokes is the ability to put things back together again—to reassemble the parts of a disassembled flashlight or toaster. Right-side damage causes such "construction apraxias" only about twice as often as comparable left-brain damage. The chances are about the same that map-reading will be impaired. Dressing apraxias, a conceptual inability to match up sleeves with arms when getting dressed, is five times more common on the right. Odds like 2 to 1, or even 5 to 1, are a far cry from language's strong 13 to 1 lateralization ratio. Thus it isn't left *or* right but both left *and* right in varying ratios for different functions.

How do we find out? Our most detailed maps come not from stroke victims but from epileptics who are awake and talking while a neurosurgeon studies their brain, prior to removing the region which is starting the seizures. But we never get the chance to study both sides of the brain at the same time in the same patient (no, contrary to the imagination of magazine illustrators, neurosurgeons do not just lift off the top of the head like a skullcap; they're very selective and make an opening overlying the problem area—on one side). Until the blood-flow and metabolic nonsurgical techniques become more sophisticated, we will continue to have difficulty determining the maps of an individual's left and right hemispheres, estimating the extent to which one side of the brain can outperform (i.e. "dominate") the other in generating a particular human ability.

But, the neurophysiologist objects, functionally everything may be all mixed up in a working brain. How can you know what the right brain is capable of by itself, on its own? Simple. As every reader of pop psychology now knows, just cut the corpus callosum which connects the left and right cerebral cortex. *Et voilà*, the split brain, isolating the two minds and all that. These studies, which won a Nobel Prize in 1981 for Roger Sperry, have caught the eye of a public that is fascinated by Dr. Jekyll and Mr. Hyde split-personality stories. Not that any of the split-brain patients have anything dramatic along that line, it's just that their right hand doesn't always approve of what their left hand is doing.

Interpreting the result is far harder than doing the surgery, as the original investigators have noted. Take, for example, the suggestion that the right brain has its own simple language center (some split-brain patients can spell out simple words with their left hand, controlled by the right brain). Many neuroscientists familiar with the childhood plasticity of the brain have worried that the split-brain patients might be highly atypical of humans in general, that some of their left-brain functions might have moved over into their right brain during childhood. That happens to young children after a severe injury to the left brain (though, as millions of aphasic adults testify, it is an ability lost with age, probably mostly confined to the preschool years), and perhaps it happens in a child with severe epilepsy. Indeed, nearly all of the split-brain patients had their epilepsy from an early age. Unlike the epileptic operated upon in the more traditional brain-mapping cases, the split-brain candidate usually has seizures originating from a whole cerebral hemisphere rather than one small piece of it. One way to try to avoid taking out the whole bad hemisphere is to just cut its connections to the good hemisphere, which was the surgical rationale for the splitting of the corpus callosum. Thus split-brain patients may be excellent candidates for studying the ability of functions to migrate from one hemisphere to the other during early childhood, rather than excellent candidates for inferring the separate abilities of the two hemispheres.

Details! Just give me the big picture—isn't that language on the left, visual-spatial imagery on the right? And isn't language all tied up with sequential things, narrowly focused cause-and-effect reasoning? And isn't the right brain better at holistic, seeing-the-big-picture sorts of things? Why not talk about two minds?

Maybe. But we must be careful to distinguish the interesting questions from the factual answers. In science, one tries to phrase questions narrowly and carefully, then gather the data so as to force nature to yield up unequivocal answers. We are just getting started at asking productive questions. We are just acquiring some of the needed tools, though they are still quite inadequate to satisfy our appetites for knowledge.

And while neuropsychology and neurophysiology struggle along trying to get firmer answers on shrinking research budgets, the left-righters are happily charging ahead full steam, prematurely assigning functions to left and right brain with all the enthusiasm of nineteenth-century phrenologists. Nowadays, one cannot even publish a drawing techniques book without invoking the liberation of the repressed right brain. I'm told that it's even a good drawing book, but its best-seller status is probably due to its gimmick of borrowing a quasi-scientific foundation. And it was a most successful gimmick at that—the taxes paid on the take from *Drawing on the Right Side of the Brain* probably exceed the world's neuropsychology research budgets.

Now perhaps that's the way to finance our research in the starvation-budgets era! We researchers can get together and write (unscientifically and anonymously, of course) a really splendid right-brain best-seller. Perhaps a marriage manual, showing how a right-brain type can achieve compatibility with a left-brain type? Or maybe a cat book for right-brain people? No, I've got it—*Cooking on the Right Side of the Brain!* We'll locate cooking skills on the right-brain map just above the right nostril. And we won't let the cook add together the ingredients in the usual sequential way—which is, of course, a left-brain way of mixing. An obvious no-no. After all, the ingredients *care* with whom they mingle, who's in the bowl already when they arrive. Instead, we'll have the cook dump all of the ingredients into the mixing bowl simultaneously, in a great splashy, holistic, right-brain moment of truth!

WHEN THINGS GO WRONG

"Anybody who thinks this is all about drugs has his head in a bag. It's a social movement, quintessentially romantic, the kind that recurs in time of real social crisis. The themes are always the same. A return to innocence. The invocation of an earlier authority and control. The mysteries of the blood. An itch for the transcendental, for purification. Right there you've got the ways that romanticism historically winds up in trouble, lends itself to authoritarianism. When the direction appears."
A San Francisco psychiatrist talking about the 1960s drug culture, quoted by Joan Didion in Slouching Towards Bethlehem

Man's destiny is to know, if only because societies with knowledge culturally dominate societies that lack it. Luddites and anti-intellectuals do not master the differential equations of thermodynamics or the biochemical cures of illness. They stay in thatched huts and die young.
Edward O. Wilson, On Human Nature

Yossarian shook his head and explained that *déjà vu* was just a momentary infinitestimal lag in the operation of two coactive sensory nerve centers that commonly functioned simultaneously. *Joseph Heller, Catch-22*

11

What to Do About Tic Douloureux

Most chronic pains
just wax and wane.
Low backs and arthritis
continue to spite us.
Try what we might,
there's no end in sight.
But the worst pain of all
has now taken the fall—
for we know what to do
about tic douloureux. *Anon.*

The worst pain in the world, say the sufferers of tic douloureux. But it is totally benign—no one literally dies from this disease, if you don't count the car accidents and suicides engendered. For it is a false alarm, though one impossible to ignore. And one of the most fascinating of medical detective stories.

But, if you must suffer from a chronic pain disorder, these sudden electric shock-like pains in the face do have a singular virtue: medical science can make them go away, permanently.

For no other chronic pain disorder—think for a moment of your friends with low back pain or arthritis—is there either a good pill or a good surgical track record. But for "tic" there is a good medical treatment—daily doses of an antiepileptic drug called carbamazepine work well in half of the sufferers—and there are two excellent surgical treatments with even better track records.

Given that perhaps 20 percent of the population suffers from a chronic pain disorder of lesser degree, there must be a lot of people around who would gladly trade their lesser but untreatable pains for this "worst" pain, given how readily it can be turned off.

These effective treatments for tic douloureux are a recent thing, payoffs of centuries of trial and error and decades of modern medical research. Benjamin Franklin's diaries from his years as ambassador to France record the primitive nerve-destroying treatments attempted 200 years ago by the leading physicians of Paris. And even today, you have to live right to get rid of your tic pain—live right close to expert medical and surgical care.

For the patients seen by the experts have usually been through years of ineffective treatment, and thus needless suffering. Some may have had most of their teeth pulled out, on the theory that a pain in the jaw that feels like a dentist's drill hitting a nerve (but continuing full-blast for many seconds) might be due to bad teeth. Many dentists cannot recognize the hallmarks that distinguish tic douloureux from more ordinary pains; besides, they frequently find dental problems anyway and are often faced with an insistent patient who demands radical treatment for an out-of-this-world pain, who will shop around until finding a dentist who will pull his teeth.

Ignorance is expensive. And in this case, excruciating. There isn't a cure for every pain problem—indeed for very few of them—but tic douloureux is now happily an exception, provided problems of education and availability can be overcome.

This litany of facts about tic douloureux, this bundle of contrasts, is only the tip of the iceberg. Neurological researchers talk of tic as one of the greatest of medical puzzles, with all of the elements of a proper detective thriller. There are dozens of tantalizing clues scattered about, some of them misleading, many exactly analogous to the classic clues of the murder mystery. Except that the story lacks a proper ending—no one yet knows for sure exactly how the dastardly deed was done. But surely such a fascinating, well-defined puzzle might, if solved, not only explain the tic pains but some other less well-defined chronic pain disorders as well.

The most important clue, appropriately enough for a good detective mystery, is the *trigger*. While tic pains occur like lightning, something usually triggers them. Shaving, or stroking a moustache, may set off a pain that lasts for many seconds. The triggering sensation is never, in itself, painful. Here is a pathophysiological version of the Sherlock Holmes dog-that-didn't-bark clue: Using a magnifying glass to focus light rays to heat up the skin will not trigger the tic pain, though causing a normal "ouch" pain. To trigger the tic pain, one must press on the skin receptors that respond to hair movement or light touch. Sometimes, just moving a single hair will be the trigger.

And here is the crime-by-remote-control gimmick: The trigger need not be in the same place as the subsequent pain is felt. Touching the moustache may trigger a bolt-of-lightning sensation felt beneath the beard along the jaw. Or perhaps over the eyebrow in other patients.

Besides these fascinating clues involving the trigger, there is the disappearing act. Tic pain is either present in great intensity, or it is absent. Unlike other chronic pains which wax and wane in intensity, tic pains are like a light switch that is either on or off. In between attacks, tic patients may be living in terror of the next visitation, but they are otherwise pain-free. The neurologist is hard-pressed to discover anything abnormal. There are typically no numb areas of the face or mouth—except as a side effect of a prior surgical treatment—and only a few patients report abnormal sensations, such as flickering, nonpainful crawling sensations here and there.

Stranger and stranger. And what is one to make of the fact that even powerful painkillers are totally ineffective against tic—yet one of the antiepileptic drugs is quite effective? Carbamazepine (trademarked as Tegretol) actually is effective in three out of four patients, but one of those three will be unable to put up with its side effects. Perhaps tic is an epileptic seizure confined to the trigeminal nerve?

Among the traits that allow the detective to locate the criminal is the *modus operandi* (does he stick to certain neighborhoods?). The other name for tic douloureux is trigeminal neuralgia, because this disease is peculiar to the trigeminal nerve's territory: tic pains are never felt outside the parts of the face served by the three branches of the trigeminal nerve; they seldom cross the boundaries into regions served by other nerves. This strongly suggests that the problem is located in the trigeminal nerve itself, not back in the brain where the trigeminal fibers rapidly become intermingled with those from other nerves. There are, however, analogues to tic douloureux that restrict themselves to other nerves of the head, such as the glossopharyngeal nerve. If anything, this rare glossopharyngeal tic is worse than the usual trigeminal version, because the trigger zone and the pain are both inside the mouth and throat. Eating or drinking may trigger the attack. Which tends to put one off one's food (remember the Garcia phenomenon from Chapter 5?).

Tic douloureux isn't rare. Just ask around in a retirement community where everyone is over 50 years old, and most people will know a sufferer, just as they will be up on the latest in cataract surgery and face lifts. While it is occasionally seen in the young, tic douloureux is commonly seen with advancing age, along with hardening of the arteries and such.

The detective story even has a false ending—a partial resolution which, like a cryptic confession in a suicide note thirty pages before the end of the book, only serves to deepen the mystery remaining. It is now apparent that most cases of tic douloureux, perhaps 90 percent, are due to the roots of the trigeminal nerve being damaged by a misplaced artery. This occurs after the three major branches of the trigeminal have merged together and are about to enter the brain stem. The neurosurgeon opens up the back of the head and uses a microscope to look down the side of the brain to the floor of the skull. Blood vessels hang here and there in the fluid-filled space between brain stem and skull, often dangling down from the overhanging cerebellum. With age, these small arteries become stiff and tend to elongate.

In patients with tic douloureux, one of these elongated arteries will be found compressing the nerve roots from the face, pounding away with each hearbeat. Sometimes it is a loop of the basilar artery coming up from below, sometimes it hits from the side, often it is a cerebellar artery dangling down from above. Where it hits the nerve roots tends to determine where the pain is felt in the face. Looking through a microscope and using long-handled instruments, the surgeon gently dissects away the filmy membranes that bind the artery to the nerve roots, freeing up the artery. Like a fire hose under pressure that has been freed from an obstruction, the artery will spring into a new position when released from the attachment to the nerve roots. This relieves a constant tension tending to pull the nerve roots away from the brain stem. Some spongelike material is inserted between artery and nerve roots to act as a shock absorber. Within a few days, the lightninglike tic pains will usually cease forever—the cure rate being about 85 percent.

This bit of delicate surgery reveals the typical cause of tic douloureux (though there are also some rarer causes such as multiple sclerosis which must be ruled out before the operation is attempted). Moving the artery effects a real cure. There is also another useful operation which, while not removing the cause, does usually stop the pains for a few years. One sticks a needle into the nerve just where it enters the skull and destroys part of the nerve with radio-frequency heating for a minute or so (just "poaching," not frying, the nerve, using temperatures of 75–80°C). While a thumb-width away from where the pounding artery is located, the site of destruction is effective because the nerve roots die all the way back into the brain. This does get rid of the pain but at the price of losing some of the nerve. In the hands of a skillful surgeon, however, about two out of three patients will lose only pain and temperature sensation in the painful region of the face—light touch will be okay but pinprick will feel dull. (Why? The small nerve fibers carrying pain and temperature sensations are more susceptible to destructive heat than the larger fibers.) Some patients prefer this operation as it is quickly done under local anesthesia and laughing gas, in an x-ray suite rather than an operating room. The artery-moving operation is a full-scale, general-anesthesia job—but it does get at the root cause, and usually leaves the nerve working properly.

So the culprit is the artery, located somewhere where it shouldn't be (similar misplaced arteries, impinging upon other cranial nerves, cause hemifacial spasm and some cases of tinnitus, vertigo, and high blood pressure). But we are still somewhere between the pseudo-conclusion and the real one, still having the problem of figuring out how the deed was done, step by step. How does the artery manage to cause a lightninglike pain? After all, the pain doesn't pulsate with the heartbeat—indeed, there is usually no sensation between tic attacks. Why antiepileptics? Why, why, why?

The mystery deepens. What mechanism will allow the moustache trigger to set off the beard pain? How will painful heat manage to avoid setting off the tic pain? Will such a step-by-step solution of the tic pain mechanism enlighten us about low back pain or perhaps epilepsy? Tune in for the next exciting episode. . . . (Provided, of course, that biomedical research survives its starvation diet.)

12

The Woodrow Wilson Story

In September, 1919, Wilson suffered a paralytic stroke which limited his future activity. After the presidency, he lived on in retirement in Washington, dying February 3, 1924.

Information Please Almanac. (*Most history books are equally uninformative about Wilson's illness.*)

The Twenty-fifth Amendment, dealing with Presidential disability, becomes part of the U.S. Constitution. It provides that the Vice-President becomes Acting President if the President declares himself disabled, or if the Vice-President and a majority of the Cabinet so declare. *Adopted February 10, 1967*

Diseases need heroes: men or women who have triumphed despite the disease. For the child with polio, one could always point to Franklin Delano Roosevelt, who campaigned on leg braces to become governor of New York and then president of the United States. For epilepsy, there is always Joan of Arc or Napoleon. The blind and deaf have Helen Keller. Woodrow Wilson provides a similarly inspiring story for both dyslexia and stroke victims—but the story of his last two years in office provides a troubling example of how brain damage can affect judgment and even block insight into one's own disabilities.

Wilson had dyslexia in childhood. Imagine not learning your letters until age 9, not reading until age 12, being a slow reader all your life. Rather than being a prescription for a life as a nonintellectual ditchdigger, this was part of the background of a man who became a professor at Princeton University and the author of a popularly acclaimed book on George Washington.

When Professor Wilson was 39, he suffered a minor stroke that left him with weakness of the right arm and hand, sensory disturbances in the tips of several fingers, and an inability to write in his usual right-handed manner. As often happens following minor strokes, there was recovery: his right-handed writing ability returned within a year.

Was his career impeded? No, in 1902 he became the president of Princeton. But the problem recurred in 1904. In 1906 it happened again, this time with blindness in the left eye (also supplied by the left internal carotid artery, which is probably where clots were originating which plugged up various small arteries in the left eye and left brain). While the right arm weakness went away, Wilson had enough damage to his left eye that he could never read with it again. Some think that his judgment was impaired in the following years—his attempts to reform Princeton academia were often impractical. By 1910 he was essentially being forced out of his presidency by the trustees.

But no matter—in 1910 Wilson was elected the governor of New Jersey. Being a university president is not the usual route to such an office (from being a zoology professor at the University of Washington, Dixie Lee Ray went on to become governor—but her stepping stones were positions as Nixon's chairman of the Atomic Energy Commission and Assistant Secretary of State, not the presidency of the university!). From the governorship, Wilson began his successful campaign for president of the United States. He won the Democratic nomination after a protracted contest, on the forty-sixth ballot.

During the campaign in 1912, Governor Wilson again suffered from mild and temporary neurological problems (now called Transient Ischemic Attacks, or TIAs, they are minor strokes without detectable lasting effects). And, a month after his inauguration, President Wilson had an episode where his *left* arm and hand were weak. All of the previous right-sided troubles had implicated the left side of the brain. Now it appeared that the right brain was also being damaged by cerebral vascular disease. But he once again recovered, an inspiration to the 2.5 million stroke victims in the U.S. who must cope with their assorted disabilities.

During his first term, President Wilson suffered from serious headaches accompanied by high blood pressure. The headaches became particularly bad at the time of the *Lusitania* sinking by a German U-boat in 1915. Were they just tension headaches, or perhaps neurological symptoms? He was re-relected to a second term in 1916, but suffered a number of TIAs during the next two years as American involvement grew in "the" world war.

Edwin A. Weinstein, the neurology professor who wrote the authoritative *Woodrow Wilson: A Medical and Psychological Biography,* also notes that President Wilson "grew more suspicious, secretive, and egocentric." An occupational hazard of the presidency—or a change in personality resulting from brain damage? The U.S. Constitution has since been amended to provide for presidential disability in office, but what neurologist would be brave enough to declare a president disabled from such a history?

If Woodrow Wilson's brain had suffered no further damage, the history of the following decades could have been very different. For Wilson in 1916 wanted Germany defeated but not crushed; he wanted Germany to be a viable member of the proposed League of Nations. He was convinced that a dictated peace "would be accepted in humiliation, under duress, at an intolerable sacrifice, and that would leave a sting, a resentment, a bitter memory upon which the terms of peace would rest, not permanently, but only as upon quicksand." The overthrow of the Kaiser in 1918 and his replacement by a democratic government raised Wilson's hopes for rehabilitating Germany. At the 1919 peace conference in Paris, he argued against French efforts to try the ex-Kaiser and to exact punitive reparations.

But then President Wilson suddenly took ill during the conference: he had vomiting, high fever, and the other signs of having caught the influenza which was sweeping Europe and later much of the world. It turned out that the virus had affected his respiratory system, heart, brain, and prostate. Indeed, judging from some of the mental symptoms (his top aide noted that, just overnight, Wilson's personality changed), Wilson may have suffered another stroke at this time or, as Dr. Weinstein suggests, have also caught the frequently associated virus of encephalitis lethargica (this is the virus whose victims often developed Parkinson's disease years later).

Even before the influenza attack, his obsession with secrecy was pronounced: none of the other American peace commissioners were privy to President Wilson's thinking. Bedridden, Wilson became obsessed with being overheard, with guarding his papers. In addition to the paranoia, he became euphoric and almost manic at times following the bedridden phase of the illness. He even became socially outgoing in ways quite uncharacteristic of the normally reticent Wilson.

But most striking was Wilson's change in attitude toward the Germans: now he himself proposed that the former Emperor be tried. Whereas he had previously insisted that the German delegates be granted full diplomatic privileges at the conference, now he was contemptuous of them. Herbert Hoover, who was there, noted the change in Wilson's behavior: before the influenza, Wilson was willing to listen to advice, was incisive, quick to grasp essentials and unhesitating in his conclusions. Afterward, he had lapses in memory, he groped for ideas, and he was obsessed with "precedents."

To ask our Twenty-Fifth Amendment question again, it seems likely that modern physicians would be able to diagnose the brain damage leading to such a personality change. They would probably recommend to their patient that he voluntarily step down. But on such evidence, would they have been able to persuade the Vice President and a majority of the Cabinet to force the President to step aside? One can imagine the discussion in the Cabinet as the neurologists tried to educate them on how brain damage can modify affect and judgment. Those not acquainted with neurologically induced personality changes would be more likely to focus on interwoven issues that they understood better—political issues such as the proper attitude toward the Germans, for example, or the allowances that must be made for people under stress.

It is hard to appreciate personality changes due to brain damage until you've seen such a patient, before and after. The first one I ever saw was a man whose head had been injured in a car accident the day before; one temporal lobe (at least) was swollen as a result of the concussion. On the door to the patient's room, the nurses had posted a sign: "Do not give this patient matches!" It wasn't that smoking was prohibited—he was mischievously lighting matches and throwing them around the room. There was nothing lethargic about this man: he was bright-eyed, aggressive with the doctors, teasing the nurses, and generally acting like a sailor in port looking for a good time. Could he have walked, it would surely have been with a swagger. Perhaps fortunately, he also had a large plaster cast on one leg; otherwise, it might have been difficult to persuade him to remain in the hospital where his brain swelling could be controlled.

One week later when again stopping in to examine all the patients on the neurosurgical service, I saw a man with an identical brain injury. This man was meek, most hesitant in his dealings with the staff, a quiet unobtrusive soul who usually averted his eyes when talking with anyone. After we left the room, I commented that the neurosurgeons' fame must have spread, that they were certainly being sent one temporal-lobe contusion patient after another. No, the attending neurosurgeon said with a smile, that man was the same patient that I had seen the previous week. I was too astonished to mind that I had fallen into a neatly laid trap which had probably been sprung upon a half-dozen insufficiently observant medical students and residents already that week. But which was his real personality? The present meek one. His family had, of course, been perplexed by the change and had told the physicians what his real personality was like. So now they knew that their patient was getting back to normal. And, *mirabile dictu,* he could also walk again—there was no longer a cast on the leg! But most cases of personality change are not this dramatic, nor can most be treated with a leg cast and diuretics. President Wilson's is a more typical case—though, because of his position, having wider ramifications.

President Wilson returned home with a treaty establishing the League of Nations. His attempts to get the U.S. Senate to ratify it were clumsy and authoritarian, not the actions of a skillful politician used to dealing with the Congress. Frustrated after five months, he decided to take his case directly to the people. Within a few days after embarking upon a speaking tour of the West in September 1919, he had developed double vision (this usually isn't either left or right hemisphere but suggests trouble in the brain stem). Wilson insisted on continuing on the speaking tour, and several weeks later, he became paralyzed on his left side: an unmistakable sign of right-hemisphere malfunction.

Another week later, after returning to the White House, he suffered a massive right-hemisphere stroke. He lost vision in the left visual field which, because of the previous trouble with the left eye, left him with vision from only one-half of one eye (this is one of those unusual sets of facts which we inflict upon medical students in a neuroanatomy quiz, to see if they can figure out that there must have been two separate problems rather than the usual one). Wilson could feel nothing on the left side of his body, besides not being able to move it voluntarily. Indeed he totally neglected the left side of his body.

Though the language functions of Wilson's left hemisphere were not affected by the right-hemisphere stroke, his voice never regained the emotional inflections and resonance of his earlier years; this aspect of speech (called prosody) is now known to be controlled predominantly by the right hemisphere.

His right-hemisphere stroke also produced a curious effect: Wilson denied that he had suffered a stroke. If you have not previously encountered the denial-of-illness syndrome, you may find this incredible. How could someone whose left body was paralyzed deny that something had happened? He indeed considered himself perfectly fit to be President (he fired his secretary of state, who had dared to call a Cabinet meeting to discuss the illness with the President's physician).

This denial-of-illness syndrome is characteristic of right-parietal-lobe damage; some patients will even deny that their left arm and leg are part of their own body. Wilson merely referred to himself as "lame." His spatial sense was disturbed: when the Secret Service took him out for a drive around town, Wilson insisted they drive very slowly and then demanded that the Secret Service chase and arrest a driver who passed them—for speeding!

Had the Constitutional amendment on Presidential disability been in effect in September 1919, Wilson's doctors should have been able to declare Wilson unfit to carry out presidential duties. But would they have done so? Like others, they could have been drawn into an elaborate cover-up to preserve presidential authority. The history of Wilson's illness gives us no comforting reassurance about how either the White House insiders or the doctors would have performed. The President's physician, Cary T. Grayson, was asked by Secretary of State Lansing to sign a certificate of disability four days after the massive stroke, but he refused. In February 1920, when the White House was issuing glowing reports on the President's health and abilities, a distinguished surgeon, Hugh H. Young, reported to the press. He said that the President had suffered only a slight impairment of his left arm and leg and that "the extreme vigor and lucidity of his mental processes had not abated in the slightest degree . . . he is in better shape than before the illness." Dr. Young summarized by saying that "you can say that the President is able-minded and able-bodied, and that he is giving splendid attention to the affairs of state."

Dr. Weinstein's excellent biography notes that at the time of Dr. Young's statement, President Wilson's left arm was useless, he could barely walk, he could not hold himself upright so as to work at a desk, he could not read more than a few lines at a time, he was subject to outbursts of temper and tears, and his periods of alertness alternated with periods of lethargy and withdrawal. And that President Wilson still insisted that he was merely lame.

Who ran the government? Gene Smith, in his book *When the Cheering Stopped,* says that the President's wife and doctors did—and that chaos and secrecy reigned. Remember, this is not a science fiction story, nor a make-believe White House thriller: this is the story of Woodrow Wilson's last two years in the White House. It not only happened, but it has only recently made it into the history books: nearly all the books on Wilson mention none of this medical history, either from an ignorance of neurology (it was just "flu" followed by a "paralytic stroke") or from the lasting effects of the cover-up conducted by Wilson's White House insiders. If the history books omit such a significant event so that we cannot learn from it, how can we avoid repeating such history?

The treaty joining the U.S. to the League of Nations was defeated in the Senate, crippling the League. Dr. Weinstein's opinion is that Wilson's stroke is what made the difference: "It is almost certain that had Wilson not been so afflicted, his political skills and his facility with language would have bridged the gap" between the two opposing sides in the Senate, much as he had done on other occasions preceding the Paris trip.

President Wilson persisted in his effort to win renomination for a third term. Pictures appeared before the 1920 Democratic convention showing Wilson in right profile (the left side of his face was paralyzed) seated at a desk holding a pen. But Wilson had no support. The Democratic party leaders prevented his name from being placed in nomination; James M. Cox and Franklin Delano Roosevelt were nominated for president and vice president. The Republican nominee, Warren Gamaliel Harding, won the election with the biggest landslide vote in recorded history. He has been described in retrospect as a "handsome and genial man, undiscriminating in his associates, lacking in political ideas or fortitude . . . totally unfitted for the presidency." American historians, when polled on who was the worst president in history, regularly select Harding.

Many feel that some consequences of Wilson's illness outlived his presidency (he completed his term of office and lived until 1924, surviving Harding) and were to be seen in the events of the following decades—in the disastrous German inflation of the 1920s during the reparations exacted by the Allies, in the ensuing reaction to social disorder which led to the rise of the genocidal Nazis, and in a second world war. All had multiple causes, but the pre-influenza Wilson anticipated many. Woodrow Wilson was a great liberal and reformer, the first world leader to fire the masses with a vision of world peace, and a courageous person who repeatedly conquered the afflictions of his chronic cerebral vascular disease. At a critical juncture in history, his brain failed him—but not obviously enough to remove him from office and let others take up the reins. During the last two years of his term following the Paris illness, Woodrow Wilson was unfit to lead the United States. He no longer had the same judgment and personality as the man whom the voters had elected.

Just as a lawyer tries to cover all the unlikely inheritance possibilities when drawing up a will, so lawmakers must try to provide for an orderly succession when the holder of a critical office is disabled—which can happen in a number of ways. Would the Twenty-Fifth Amendment, which seems so inadequate to deal with Wilson's earlier problems, have covered the final Wilson tragedy?

Even if neurologists could diagnose a serious change in perceptual abilities or in personality, could they convince the President to voluntarily step aside? What happens when a strong-willed President's judgment, like Wilson's, is clouded by his illness? Judging from the difficulty that physicians have in persuading ordinary patients with right-parietal-lobe damage that they are ill, the physicians would probably have been rebuffed. Could they then convince the Vice President and a majority of the Cabinet to notify Congress that the President was disabled? The Twenty-Fifth Amendment seems to assume that either a President will be rational enough to declare disability personally, or that the President will be in coma, unable to interfere in the Cabinet's decision. Suppose that, like Wilson, a President were to fire the questioning Cabinet members first? *The Twenty-Fifth Amendment would seem not to cover the most serious and most prolonged Presidential disability yet encountered in more than two centuries and forty Presidents.* It remains to be seen if the Constitution's disability provisions function any better than those of the Divine Right of Kings (which allowed George III—the *bête noire* of the American colonists—to rule England for many decades while insane on and off, even confined to a straightjacket at times).

Neurology was established as a medical specialty in the nineteenth century by a series of great physicians-investigators, but the recognition of subtle intellectual deficits in stroke patients was only beginning in Wilson's time. Because such "higher functions" cannot be studied easily in experimental animals, progress has been slow in comparison to other areas of brain research. In 1920 a remarkable era began, during which the individual nerve cells have been explored, the reflexes extensively studied, great inroads made into understanding the functions of sensory and motor systems, and many specialized cortical areas identified. We now know more about developmental dyslexia, from which Wilson initially suffered, and about recovery of function after strokes, which permitted Wilson to recover from his many earlier strokes so successfully. We now have diagnostic techniques such as computerized tomographic (CT) scans, which would have detected much of Wilson's brain damage, and therapeutic techniques such as vascular surgery, which, if performed at age 39 after Wilson's first stroke, might have cleaned out the arterial lining problems in the carotids which probably formed the clots.

Yet we still lack a body of reliable physiological and anatomical facts with which to understand personality change and denial of illness. One must rely more on the art of the experienced physician in such cases, not on the hard facts of science dispensed by machines. But it is not clear whether even the most expert of modern physicians would be able to protect the world from the consequences of a similar brain malfunction in a modern president.

13

Thinking Clearly About Schizophrenia

If schizophrenia is a myth, it is a myth with a strong genetic component. *Seymour Kety, in reply to Thomas Szasz*

Ask most people for a definition of schizophrenia and they'll say something about split personality. Indeed, the word has entered popular usage: "I feel rather schizophrenic on that subject," meaning that you have two different opinions on the matter. Even the etymology suggests it: it comes from the two Greek roots meaning "to split" and "mind or heart." But two plus two isn't always four in the history of words, and certainly not in science, where our understanding improves with time but the names remain the same.

Ask someone in the mental health professions for a capsule description of schizophrenia and you'll get an entirely different response: "It's a thought disorder. The person often hears voices commanding him, or giving a running commentary on his actions, or perhaps hears his private thoughts being spoken aloud (even though they aren't). They have trouble maintaining a train of thought, often jumping to a totally irrelevant topic. Just as in the hard-of-hearing, a feeling of being persecuted or spied upon is very common in the schizophrenic." People who have suffered the disease for a long time are more likely to be withdrawn, not know what day it is, not say very much, be intellectually impaired, not have normal emotional responses to disturbing news." Schizophrenics certainly have trouble thinking clearly, but that is no excuse for the rest of us thinking fuzzily about schizophrenia. But views of schizophrenia have varied from T. Szasz's "it doesn't exist" and R. D. Laing's "it is a sane response to an insane world" to notions that it is a biochemical disorder like diabetes.

The split-personality notion of schizophrenia is really a description of a very rare personality disorder which has been seized upon by a public unfailingly entertained by Dr. Jekyll and Mr. Hyde. In terms of statistics, one can essentially forget about it—but one cannot forget about the 1 percent of the population that becomes schizophrenic, usually as young adults. Nor, apparently, can the English language do without a word to serve the role which the misnomer, schizophrenic, has been serving.

To tackle the easy problem first, there is a perfectly good word that will fill the bill: *chimeric.* A chimera, just in case you have forgotten Greek Mythology 101, is a she-monster with the head of a lion, the body of a goat, and the tail of a snake (it was slain by Bellerophon astride the winged horse Pegasus). The word has found a home in at least two aspects of the scientific literature: genetics and cognitive psychology. Genetic engineering techniques make it possible to mix the genes from two different animals, provided that it is done early enough in development. One can wind up with, for example, a fly that is a patchwork mosaic of colors and other surface properties which illustrate different "compartments" in development under control by different precursor cells. Since some of those cells belonged to one embryo and some to another, the resulting animal is an admixture. An individual that is composed of parts of several individuals is called a chimera. The only humans who might qualify are the recipients of transplanted kidneys and such.

Cognitive psychologists aren't studying such chimeric humans; they are studying how normal humans react to artificial chimeras. They have demonstrated that, when you look at a picture of a person's face, it is the right side of that face which impresses you most strongly as regards emotion. The psychologists take pictures of an actor acting out various emotions: happiness, rage, sorrow, disgust, and so forth. Then they take a razor blade to each picture and cut it down the middle of the face. Thus they can paste together a right face which is happy with a left face which is sad: a chimera. Such a picture is briefly flashed on a screen—and the subject reports that it was a happy face. The right side of the face was in the subject's left visual field, which reports first to the subject's right brain, which has an interesting cortical region specialized for judging the emotions on other people's faces (discussed back in Chapter 5).

Thus, the next time you are tempted to describe yourself as "schizophrenic" when you are torn between two different opinions, don't forget that someone may take you literally and suspect that you are experiencing hallucinations. Instead, call yourself "chimeric." It may, however, be the better part of valor to revert to "schizophrenic" should you see someone approaching on a winged horse.

In which case, of course, you may actually be schizophrenic.

> Psychiatrist: Why do you flail your arms around like that?
> Patient: To keep the wild elephants at bay.
> Psychiatrist: But there aren't any wild elephants here.
> Patient: That's right. Effective, isn't it?

This patient not only suffers from visual hallucinations but also from what is called "knight's move" thinking—linking concepts whose connection requires a novel dogleg leap of the imagination, such as the cause-and-effect of arm waving and no elephants. This sideways leap need not be that absurd. Some would explain the allegedly increased numbers of artists and scientists who suffer schizophrenia by saying that a small amount of this trait aids creativity, enhancing the ability to perceive novel interrelationships, to break free of the bonds of straightjacket thinking. So that pre-schizophrenics might become employed in creative jobs in disproportionate numbers to their usual 1 percent prevalence in the population as a whole. Full-blown schizophrenia, however, can be so disorganizing that serious creative work becomes impossible; one can no longer sort out the metaphorical wheat from the chaff, nor communicate understandings to others.

But auditory hallucinations are much more common in schizophrenia than visual ones. People who hear voices, realizing that others do not also hear them, quite reasonably rub their ears to make sure that an earphone isn't there. That possibility eliminated, they may wonder if some secret technology is at work, bypassing their ears and directly inserting the voices into their head as if it were a radio receiver. The voices can seem far more real than in a dream, and suspecting a secret CIA technology is more logical than trying to deny that the voices exist.

I have actually had schizophrenics seek me out for an expert opinion on this subject, on the suspicion that their psychiatrist wasn't up on the latest in neurophysiological techniques. It is not easy to convince them that such broadcasting into brains is presently technically impossible, because their personal experience tells them otherwise. It is easier to believe in secret CIA methods than to deny the evidence presented by your own brain. It is easier to believe that the CIA has kept the method secret even from the neurophysiologists. However, the technology simply doesn't exist. A method of inserting voices directly into the brain (or of reading out detailed information from a person's brain) would have revolutionized neurophysiology research if it existed, and one cannot successfully keep hot new technologies like that a secret from the professional researchers in the field. It would take many decades of gradually improving techniques to lead up to such a technology; it would take a revolution in our knowledge about hearing and speech; and it just hasn't happened.

Persons experiencing such voices usually do not know enough about the brain to appreciate the most likely explanation: that the mechanism which the brain uses to distinguish between incoming sensory messages, and vivid memories of previous experiences is temporarily not working well. The hallucinations, just like ordinary nighttime dreams, are recalled information from the person's memories—though often pieced together so creatively that they may seem new. They may try to tune out the voices by suppressing auditory input, which may be why they suffer the same feelings of paranoia as some people whose hearing is impaired. If the voices are bothersome, they can usually be quieted by antipsychotic drugs. A friend of mine, when in his late twenties, was hospitalized after hearing voices commanding him to take off all his clothes and perform pushups in the middle of a downtown street. He followed orders. The antipsychotic drugs had him back to work within weeks.

Effective drug treatment for schizophrenia is only about thirty years old. A French naval surgeon, H. Laborit, was treating sailors with worm infections, no less. He noticed that his favorite antiworm medicine, chlorpromazine, also had a calming effect in mentally disturbed sailors with worms. In the fifteen years that followed, many related drugs were tried out by many investigators. What the effective ones all had in common was that they blocked the actions of a neurotransmitter known as dopamine. Now, after even more experience, we know that it isn't just dopamine.

Schizophrenia is a set of symptoms; like epilepsy and the common cold, it is not just one simple disease. So it is not surprising that the antischizophrenic drugs do not have uniformly good results. Yet which drugs work on what symptoms tells us something about the neurochemistry of schizophrenia. Nothing much seems to work on the social withdrawal and the flat emotions of some schizophrenics, though there are some promising new leads.

Dopamine is a neurotransmitter in common use in many areas of the brain; too much of it (or supersensitivity to a normal amount) appears to be responsible for some schizophrenia symptoms such as stereotyped behaviors. Indeed, you can induce such schizophrenialike symptoms with drugs that enhance dopamine; it is a common side effect of levodopa, used to treat Parkinson's disease. My elderly landlady in Jerusalem started hearing voices after accidentally taking a double dose of her medication for Parkinsonism. Overdoses of amphetamine (which are thought to stimulate dopamine release) induce a schizophreniclike psychosis, one which is readily overcome by the usual antischizophrenic drugs; this provides a way of inducing "schizophrenia" in laboratory animals, an all-important step in furthering medical research.

Other symptoms, such as withdrawal and autistic behaviors, flatness of affect, and lack of motivation, have been blamed upon disorders of another neurotransmitter, norepinephrine, which is manufactured from dopamine. Some research groups think that the enzyme that controls the conversion of dopamine into "nor-epi" may be at fault.

Another line of investigation notes that people with schizophrenia do not develop arthritis, and vice versa. Furthermore, schizophrenics are often resistant to pain. And they improve mentally when running a fever (a fact first noted by Hippocrates). It is even said that people with both schizophrenia and epilepsy have an unusual alternation between the symptoms of the two diseases, with seizures being rare during periods of madness. Whatever can the medical detective make of these fascinating clues?

David Horrobin, a Montreal endocrinology researcher, notes that all these facts could be explained if prolactin, a hormone secreted by the pituitary gland at the base of the brain, protected us against schizophrenia. Prolactin, via stimulating the manufacture of fatty acids called prostaglandins, is all tied up with arthritis, pain, fevers, and seizures.

- Rheumatoid arthritis is a disease where there is excess production of prostaglandins.
- Prostaglandins play a key role in inflammation and pain.
- Prostaglandin levels rise during fevers and seizures.

And it explains the therapeutic effects of the common antischizophrenia drugs that block dopamine: they enhance prolactin output, so there are more prostaglandins around to do their antischizophrenic thing. Whatever that turns out to be—maybe the enzyme mentioned earlier?

Such research proceeds slowly; U.S. biomedical research dollars have been constant since 1974 so, corrected for inflation, our basic research efforts have been steadily shrinking. Considering the enormous social costs of mental illness, society's lack of investment in research seems both stupid and inhumane. I was just going to say shortsighted, but stupid does seem the right word.

Schizophrenia is especially a disease of the young adult—people with their whole life ahead of them. It lays them low, and in addition to the personal and family tragedy, if often disables them at great cost to society. After puberty, case after case appears until, by middle age, 1 percent of the population has been affected, uniformly across social classes. Initial hospitalizations for men peak at about age 18, with many cases first occurring in the twenties. For women, the cases peak at about age 30. Few cases first appear when people are in their forties (for severe depressions, in contrast, first hospitalizations peak at age 55).

How many friends do you have from high school who were hospitalized for mental illness as young adults? Given the percentages, you should know a few. Two of my high school friends died in their twenties; both were very gifted socially and academically, one serving as the president of my senior class. They were identical twins, and when one identical twin gets schizophrenia, the chance of the unaffected twin eventually coming down with schizophrenia is about fifty-fifty. Until I heard the news about my twin friends at my class reunion, that had always just been one of those impersonal statistics that I carried around in my head.

For fraternal twins and other siblings in general, the chances are about one in six of coming down with schizophrenia if a sibling is affected. Twins aren't any more likely to get schizophrenia than anyone else, unless they have close relatives with the disease. It's just that a monozygotic twin has a closer relative than anyone else has. The "concordance" holds even if twins are raised apart, when one's illness could not affect the environment of the other (sixteen such pairs of separated twins with schizophrenia have now been discovered and studied).

That strongly suggests a genetic predisposition, that schizophrenia is inherited (though it is not via simple Mendelian genetics, unfortunately). And there are now many findings that demonstrate a biological basis, the latest being a valuable new series of studies coming out on the brain anatomy of schizophrenics, thanks to CT scans which allow cross-sectional views of living brains via computerized x-rays. The fluid-filled areas are sometimes a little bigger than normal, and characteristic asymmetries in the left brain compared to the right brain are sometimes absent in schizophrenics (they are more symmetric). Not all schizophrenics show the nonstandard anatomy, but many do. Timothy J. Crow, a British psychiatric researcher, proposes that there are two major subdivisions of schizophrenia: that the negative symptoms (flatness of affect, intellectual deterioration, poverty of speech) accompany the anatomical changes and are not affected by drug treatments; and that the positive symptoms (hallucinations, illusions, thought disorder) reflect a neurohumoral component to the disorder that is treatable with drugs.

And indeed the twin studies also demonstrate that environment has a role in schizophrenia. Since monozygotic twins have identical genes, how else can you explain one twin developing schizophrenia but the other one not? That is, after all, the other side of the fifty-fifty statistic: half of the twins with a schizophrenic twin do not themselves come down with schizophrenia. So environment too must play a role in the disease. But it has been an elusive factor to identify, compared to the biological ones.

What follows is a story that I'd like to forget, like a bad dream—but it pointedly illustrates what is known as stereotyped behavior and flat affect. If you wondered about that bit of technical jargon earlier, and have a strong stomach, read on. But don't blame me if you can't sleep tonight. You can always skip the next page.

I was once the foreman of a jury in a first-degree murder trial, where the defendant did not contest the facts but argued insanity under the classical McNaghten rule. He was 20, with a history of mental problems, poorly treated (this was back in the old days before community mental health clinics became common; I suspect he never saw a psychiatrist until he landed in jail). Weeks earlier, he had choked his girl friend to the point of unconsciousness, but no one had done anything about the warning signal (girl friend included, which caused some jurors to comment upon *her* sanity).

Jobless, he had begun hanging around a neighborhood where he had once lived as a child. In one house there lived a deaf woman, 76, who always worried about being murdered for her money, an attitude the defendant had probably absorbed while growing up nearby. Though the run-down condition of her house would not have suggested more income than a social security check to anyone sane, the defendant broke into it in the middle of the night and beat the poor old woman to death in a frenzy. She was stabbed thirty-five times with a long knife and sixty-seven times with a metal bedpost. Thereafter, the defendant burglarized the house across the street, skillfully emptying the pockets of a sleeping man. Then, at seven in the morning, he hitched a ride downtown from a neighbor who knew him, who politely asked what he was doing back in the old neighborhood.

He was not walking about in a daze, totally incompetent—which caused the jurors no end of trouble in deliberating the insanity defense. But the flatness of his emotions got through to us: How could any sane person show such lack of reaction to a frenzy of brutal violence as to be cool enough to go pick someone's pocket within the following hour? Only a dream, forgotten upon awakening, could allow the diassociation of such emotions. And in schizophrenia, dreams and reality can often become confused. The inhibitions of muscle movement which we have during dreaming usually keep us from acting out our dreams. But in some schizophrenics, reality can be insecure, it can come and go, intermixed with what are only dreams—but which are sometimes acted out. (Let me hasten to emphasize that only a few schizophrenics are dangerous to others.)

The defendant's facial expressions had been consistently "weird," as the arresting police officer described them (he was arrested while returning to the house a week later to view the body). He smiled a silly, inappropriate smile (a stereotyped behavior), even when sad (flat affect). One judge had been so impressed by that weird smile that he decided to ignore the current medical opinion which said that the defendant was now sane enough to help his attorney. So the judge delayed the trial again. Finally, three years after the murder, physicians and lawyers and judges agreed that he was capable of assisting his attorney in his own defense—though everyone seemingly agreed that he was probably still medically psychotic.

We poor jurors had to keep straight three different definitions of insanity: (1) the criteria for being able to stand trial now; (2) the criminal insanity criterion: "not responsible for his actions at the time"; and (3) the medical definition of psychosis whose elements I have outlined above. Our job was to make a judgment about the second criterion, and we eventually decided that he was not guilty by that criterion. Though we were charged to judge the past, some jurors may also have had unspoken thoughts about the future: that surely the defendant would be better off in a treatment setting such as a state mental hospital than in a punitive warehouse setting such as a state prison (and eligible for parole in eight years). I am sure, however, that no one had any illusions about whether the defendant was safe to mingle in society.

Jurors seldom get much feedback about how things turn out. By happenstance, I did. The state mental institutions had found this psychotic young man too dangerous to the other residents, and lacking proper facilities for such high-risk troubled people, they simply transferred him to the state's maximum-security prison.

When I was the state president of the American Civil Liberties Union, one of our staff attorneys came back from a visit to that prison telling of a really spooky guy, quite unlike the other prisoners, who had cornered him while he was interviewing inmates for a class-action lawsuit on poor medical care. I surprised the attorney by guessing the inmate's name. I then wrote to the judge about the quality of the medical care that this not-guilty defendant was receiving—and in which kind of state institution, at that. The judge investigated and the institutional officials explained their problem with inadequate facilities, record populations, and budget cuts. The defendant-patient-prisoner did find one way of getting attention: he escaped (and was recaptured).

Not all medical problems are treatable (and his, admittedly, were not encouraging), but there is no excuse for compounded neglect. Not only are the taxpayers' representatives reluctant to invest money in medical research ("it's someone else's responsibility") but they often simply ignore people for whom they have direct responsibility. If they were parents failing to look after their children, they would be hauled into court for their willful neglect and subjected as well to the attentions of press and media. Yet they, and the taxpayers themselves, fail in their responsibilities and get away with it in the name of holding down taxes. We do not suffer from lack of resources, but we certainly have a problem in setting priorities for their use. It says something about our society that we spend many times more money on tobacco and cosmetics than on biomedical research. And more on pet food than on rehabilitation.

Despite my week-long education on insanity defenses, I still don't understand them; they certainly bear little relationship to the body of knowledge represented by modern biological psychiatry. Nor do I have a revolutionary new system to propose for relieving the confusion they engender in our courts. So I have a lot of empathy for the newspaper reader who cannot understand insanity pleas. It is indeed a complicated business, and not just because human behavior and schizophrenia are so complicated. It is not easily encapsulated in those little bite-size tidbits favored by newspapers and television news when they need something to fill in between ads; you need to approach the legal policy on insanity more the way a dog tackles a bone, by chewing on the subject for a while from one direction, and then from another angle.

The first step is to be suspicious of glib generalities, such as those uttered in outrage over the not-guilty-insane verdicts for presidential would-be assassins. Back when I was a beginniner on this subject myself, I got to see how eleven other ordinary nonexpert citizens approached the criminal insanity issue. I suspect that most of the people who write indignant letters to the editor would probably, after spending a week locked up in a jury room facing such issues themselves, see things much as average jurors do. It is surely a sobering experience. And such considered sober opinion from a jury of peers is where most of those not-guilty-insane verdicts originate.

Schizophrenia is not a myth but a very real disease, distressing to its sufferers and to society. The real reason why we have such trouble talking about it realistically, however, probably has little to do with confusing the disease with that suffered by Dr. Jekyll (or was it Mr. Hyde?). Schizophrenia is a disorder of thought, of our perception of reality. And we have little idea how our brains work to create thought, how we form a mental picture of reality and test it against the facts. We don't even understand dreams: how we can inhibit our actions during them, how we readily forget them unless we replay them immediately after awakening. Until basic neurophysiology gets a better grasp on such subjects, until it translates such an understanding into a form readily comprehended by jurors and newspaper readers, our reactions to schizophrenia are going to continue to be—well, chimeric.

14

Of Cancer Pain, Magic Bullets, and Humor

In another country, I attended a riot out of that curiosity which is not only known to kill the cat but on occasions the man. I was a passive observer on the pavement between rioters and paratroopers, well to one side, talking with a reporter. Suddenly I felt a dull thud on my forehead—a glancing hit had ripped a jagged 4 cm-long slit. For an hour, the injury was totally painless. I began to worry that my family would be worried if they heard that I was injured and later the thought crossed my mind that infection might result. At this stage, I began to feel a minor deep dull headache.

By the next morning, deep and superficial pain and tenderness were very obviously present, distressing and disturbing. This died down day by day as the wound healed cleanly without problems. In spite of this obviously impending successful end of the episode, and in spite of the good company and interesting work, I found myself unusually listless, sleepy, irritable, unable to concentrate, shunning food and company. This also disappeared over a period of 7–10 days. *Patrick D. Wall*

Pain is about as pleasant as a fire alarm—and also as useful. It's nice that your leg nerves are wired up in the spinal cord so that stepping on a thumbtack results in an automatic withdrawal reflex that limits the damage, with the foot lifting and the other leg straightening to take the weight. If pain has a bad name, it is because it is sometimes like a fire alarm that can't be shut off—continuing insistently for a long time after it has alerted you that something is awry.

If you have broken a bone, such continuing pain probably serves a useful purpose. You become reluctant to move the affected part of your body, and this immobilization aids the healing process. In the days before plaster casts when the human nervous system was being shaped by evolutionary pressures, such continuing pain made some sense. Injured animals hole up somewhere while they heal. The late lassitude that Professor Wall experienced may be useful in protecting the healing wounds; similarly, the early pain was merely sufficient to get him to seek care, but not severe enough to immobilize him at the site of the injury (indeed, 37 percent of emergency-room patients report that they did *not* feel pain until some time after the injury). In looking at some of the disorders of pain behavior, we tend to seek their origins in these normal phases following injury and in how they vary from one person to another.

There is a lot of needless suffering associated with chronic pain, suffering that serves no useful purpose whatever. Evolution has evidently not worked very well at preventing this, though our scientific ability to understand the pain mechanisms may eventually accomplish that goal. Minimizing suffering is a human, humane goal but apparently not an imperative of evolution.

The progress report for this humane goal is, however, not one of the success stories of our civilization. We are just starting to recognize that the brain's mechanisms are quite different for the thumbtack pain and the months-long pain, just starting to understand some of the relationships between chronic pain and depression, just starting to develop some tools for sorting out the problems in animal research which have prevented progress in the past.

Morphine and similar drugs work very well when you break a leg. The problems come when the physician tries to treat chronic pains such as those from neuralgias or headaches or arthritis. It isn't merely the addiction problem with morphine. It's that all the strategies used for dealing with acute pains—bed rest, immobilization, potent narcotics of various sorts—don't work for chronic pains. They often make them worse.

One of the most important decisions that a physician must make is whether the patient's pain falls into the acute category or the chronic one, because the treatment of chronic pain is usually almost the *opposite* of that for acute pain. Anything stronger than aspirin is usually avoided. The patient must not hole up like the injured deer with the broken bone. Walking around the block may seem unnatural to us when we are in pain (as our evolutionary heritage in such matters is much like the deer's; we act much as Professor Wall did a week after his injury), but activity must be forced. If the chronic pain is complicated by depression, treating the depression with antidepressant drugs and psychotherapy may help relieve the pain.

So the natural instincts for dealing with chronic pain must be resisted, again proving that evolution is imperfect. But what really works well? While there are some exceptions like tic douloureux and its very specific medical and surgical therapies discussed in Chapter 11, the therapy of chronic pain is a mixed bag of psychological, medical, and surgical tricks. It is often not done well at all. The patients who filter through to the pain clinics have typically been through a bewildering array of medicines and operations, have literally gone from one doctor to another, tried everything.

Pain clinics are a recent invention, though their antecedents go back to World War II. My colleague John Bonica founded one in 1961 at the University of Washington in Seattle; during the 1970s many pain clinics sprang up at both university hospitals and private clinics throughout the world. It was an idea whose time had come, but there is no revolutionary technique at its core. Rather it serves to pool the knowledge and techniques of a number of medical and psychological specialists, serves to create full-time specialists in pain therapy among the anesthesiologists, serves to better identify the conditions for which surgery offers effective therapy, and serves to create a practical body of knowledge about what works and what doesn't.

A proper reductionistic approach seems at the very core of "scientific medicine": one seeks to get at the root causes, not just to try to make the symptoms go away. On this theory, one treats the pain by eliminating its cause. Granted, some symptomatic treatments make about as much sense as using makeup to cover the jaundiced skin color of hepatitis. But sometimes one must step back a minute and consider whether, in human terms, it makes sense to put all your eggs in one basket. There will always be diseases about which we can do nothing; we will all die of something and would prefer to die pain-free and with dignity. Furthermore, there will always be irreversible history—accidental injuries to the nervous system that cannot be undone, scars that cause pain and cannot be removed without making another scar.

Particularly in chronic pain, the symptom is often more of a problem than the malfunction giving rise to it. The minor alterations in nerves that give rise to neuralgias are not, in themselves, worrisome. The irritations of blood vessels and the brain's covering membranes giving rise to most headaches are benign. The blood vessel that traumatizes the trigeminal nerve in tic douloureux is usually not preventing the nerve from conducting normal nerve impulses; it infrequently causes serious damage. Most disk disease does not threaten the health of the spinal nerves and the spinal cord (though the exceptions can be quite serious indeed). Most arthritis pain is all out of proportion to the need to warn us not to overuse the joint. But the pains associated with such physically minor irritations can disable even the most stoic person, can even lead them to contemplate suicide.

In the case of cancer pain, the situation is particularly absurd. In evolutionary terms, the pain serves no useful purpose at all. Even in these modern times when something can often be done for the tumor, the pain does not serve a useful warning purpose, to get someone to see a doctor: in most cases, pain is not an early symptom of cancer. It is not even inevitable; many cancer patients never experience serious pain. Yet pain is probably the most feared aspect of cancer, and for good reason too—tumors that invade bone or peripheral nerves or pancreas are especially bad news.

You might think that, because it is such a widespread problem (perhaps a half-million sufferers in the U.S. at any one time), physicians would be well trained in managing cancer pain. The medical books on oncology are thick tomes, and one might expect to find therein several long chapters on pain problems and their management. But many do not mention the subject at all; others have only a brief mention. One count showed that the number of pages on pain problems constitute less than 0.2 percent of the cancer textbooks. The knowledge base is not as poor as that, but it is often taught poorly, having fallen between the cracks of our medical specialties.

A similar impression occurs when one looks at the budget for the U.S. government's arm for cancer research, the National Cancer Institute. An analysis of its program in recent years reveals that of its billion-dollar budgets, only 0.022 percent went toward cancer pain research. Cancer pain researchers are surprised when their grants are assigned to the NCI's Community Relations office for administration—until they look at the NCI's table of organization and can find no more logical home. It might be argued, with some fairness, that few good research proposals are submitted in the area of cancer pain research, but it is also obvious that cancer pain research is an orphan area, overshadowed by the noble quest for the magic bullet that will cure cancer. And hence avoid cancer pain. But our natural tendencies to seek root causes and cures perhaps need to be tempered—hedging our bets a bit better than the just-quoted percentage. Besides, our knowledge about how to treat cancer pain might carry over into other noncancer diseases—such as what will replace cancer near the top of the fatal-disease list should that magic bullet ever appear, if the cell growth or immunology researchers strike it rich.

While much cancer pain is probably due to tissue damage (and is thus "proper" pain rather than "false alarm" pain), there are many clues that cancer also alters the nerves, mangling the messages. There is a form of muscular weakness (known as the Eaton-Lambert syndrome) which may appear at a site distant from a tumor, as when a lung tumor causes hand weakness (nerve endings no longer release as much neurotransmitter in response to an impulse as they normally would, just as in some forms of myasthenia gravis). The pressure from tumors sometimes injures a nerve, causing it to sprout and attempt to regrow—during which time it will be abnormally sensitive to movements and various things circulating in the bloodstream. Sometimes tumors grow along nerves, completely changing the environment of the nerve and making it malfunction. My own pet (but unproven) theory is that much cancer pain is very similar in mechanism to neuralgia pains—that if we learn to prevent one, we will have gained a handle with which to manipulate the other, that cancer pain research will prove relevant to garden-variety low back pains (no pun intended) and vice versa.

If there is one overwhelming lesson from basic biomedical research, it is that most problems cannot be effectively tackled head-on. You have to find out a lot about many related things and then, some lucky day, a path will open up toward one of those problems easily identified by the public as a major concern. This lesson, as well as the scientific tendency to seek root causes, means that we properly put most of our eggs in a basic research basket. In the case of pain research, this has meant studies of the specialized sense organs embedded in skin, muscle, and viscera. It has meant tracking the pathways taken by pain and other sensations through the spinal cord and brain. It has meant investigating the normal mechanisms which the brain uses to regulate the sensations reaching consciousness, its ways of turning off unwanted messages.

Many of the neurobiologists I know would bet that the study of peptide hormones (mostly done in moths, snails, and the like) will prove quite relevant to pain research: one of the brain's control mechanisms involves enkephalin. It is a short-chain protein, a mere five amino acids long: tyrosine, glycine, another glycine, phenylalanine, and finally leucine (or sometimes methionine). Some think that it or a close relative may prove to be the brain's own version of morphine, perhaps even a neurobiological "magic bullet" for pain. Certainly placebos suggest that the brain itself has a mechanism for raising pain thresholds in a morphinelike way: in perhaps 30 percent of patients, a sugar pill taken under the impression that it is a potent painkiller seems to work as advertised. And, lest one think that it is all "psychological," consider this: An antimorphine drug (naloxone) will make the placebo become ineffective, this return of the pain suggesting that the placebo stirred up a morphinelike hormone or neurotransmitter in the brain (naloxone blocks morphinelike receptor sites). The psychological set seems to tap into a normal brain mechanism for regulating pain sensations, at least for a week or two.

(Placebos raise a problem in these days of the pharmacist labeling pill bottles with their contents. One cannot admit that the pill is nothing but sugar if it is to work, so a fancy brand name is needed. Among the proposals made in the scientist's humor magazine *The Journal of Irreproducible Results* for what to name a brand-name placebo are Confabulase, Gratifycin, Deludium, Hoaxacillin, Dammitol, Placebic Acid, and my favorite, Panacease.)

Most research requires intense specialization in order to make progress. Thus most neurobiologists are not familiar with the differences between acute and chronic pain (of course, neither are most of the physicians treating pain). To counter the natural tendencies to a narrower and narrower focus, there have been a series of interdisciplinary research ventures in the last decade. Groups of researchers from different backgrounds join together, motivated by a common interest in a particular disease. For example, a pain group might include neurophysiologists working on how nerve impulses get started, on how pain messages are handled in the brain, on how pain affects sleep; it might include neuroanatomists tracing the pain pathways in the spinal cord, neuropharmacologists studying how known pain-relieving drugs act upon nerve cells, behavioral neuroscientists trying to modify pain in rats. And besides the basic scientists, there would be the anesthesiologists, psychologists, psychiatrists, and neurosurgeons treating pain and pursuing their own research interests, perhaps even sociologists studying how the news about improved cancer treatments spreads on the medical grapevine. The clinicians and basic scientists may sometimes work together, perhaps looking at cancer patients and wiretapping a nerve, listening in on the messages being sent from the tumor to the spinal cord and brain, trying to see the tell-tale clues of false alarm messages, trying to figure out which "proper pain" messages are being sent.

Such ventures have their problems, as each scientist may feel somewhat isolated from the parent discipline or feel out of the mainstream of his own field's research thrust into ever deeper layers of reductionism. The ventures are also financially unstable, usually completely dependent upon government grant money doled out for limited two- or three-year periods—at the end of which time the whole venture can fall apart for lack of jobs (the universities, in which most such interdisciplinary ventures are housed, rarely try to guarantee such jobs with tenure, thus creating an academic version of second-class citizens).

The clinicians, should they have to give up their research activities, are at least comforted by being able to double their academic incomes by going into private practice. Unlike the clinicians, the Ph.D.-trained basic scientists cannot fall back on treating patients; they are in a very exposed, precarious position which periodically causes them to doubt their own rationality. A friend of mine, who trained with several Nobel-class people in neurobiology and whose published work is well known, is now making his living by selling computers to small businesses that want to automate their bookkeeping. Two other postdoctoral researchers used to support themselves by applying their mathematical knowledge of game theory (remember "tit for tat"?) to card counting, and would make semiannual trips to Las Vegas to cash in on their computer-pretested schemes at the blackjack tables, regularly earning enough money to support themselves for a little longer (in the usual poor graduate-student style, nothing fancy) while they did neurobiology research. (I can tell this story now only because they too have given up on such a chancy existence in favor of well-paid jobs setting up computer installations.) Another talented researcher sells real estate for half the year, then commutes 5000 kilometers to spend the next six months doing neurobiology again. But living on a shoestring gets a little tiresome after a while, especially if there are children to eventually send to college. Unfortunately, the clinician or the basic scientist who leaves research is, for a variety of reasons (such as having to lower their standard of living), quite unlikely to ever return. And so the fits-and-starts of federal funding for biomedical research regularly cause highly trained people to be lost forever to research, however useful their second careers may turn out to be.

It makes no sense at all—but then, neither does the overall situation. Judging from the low level of investment in research and development, you'd think that biomedicine was a slowly changing industry like an electrical utility. But utility executives probably invest far more of their budgets in R and D than does the health care "industry." The main problem is that, due to the small-business nature of most health care practices, there is very little investment in research except via governments and taxes. One way of tying research expenditures to the overall health care budget would be to tax health insurance premiums to create a biomedical research trust fund, in much the way that the interstate highway system was built through a gasoline tax creating the highway trust funds. That way research would at least keep pace with inflation; it's been downhill since about 1967 as one president after another has proudly announced a 3 percent increase in the biomedical research budget—in a year with 10 percent inflation. The medical definition of starvation is when the patient continuously loses weight, with no end in sight—which is why we talk about "starvation diets" for biomedical research. It isn't hyperbole: the funds available are absurdly small by business standards for R and D, and they keep declining despite the self-congratulatory press releases.

On the bright side, it might be said that the psychology of humor may prove relevant to chronic pain. Scientists may not headline a research grant as a study of humor (for fear of losing a year of their time coping with a press release from Senator You-know-who), but there are a variety of clues which scientists are convinced need pursuit. The endorphin and enkephalin researchers are looking to see if humor stimulates natural pain defenses. Then there is laugh therapy, such as Norman Cousins (the editor of the *Saturday Review*) utilized in coping with a serious illness.

The other line of thought comes from cognitive psychologists, who talk of the schemata or templates in the brain, to which we are always trying to match up the situations in which we find ourselves. In the days before cognitive psychology, a prominent British neurobiologist, J. Z. Young, expressed it this way:

> We laugh together when we agree, when we understand the same rules. . . . One of the ways of emphasizing a common use of rules is to laugh at someone who does not

> share them—hence we laugh at mistakes, odd men out, freaks and Ugly Ducklings generally. . . . [Even in science, researchers exploring new avenues, the doubters] who are trying to find new rules, are always ridiculous, because they are not satisfied with the old ones. Jokes have the same origin, the fact that one sees the disguised or oblique point gives the assurance of understanding the rules of ability to communicate. Jokes about sex are the funniest of all, because they show an agreement about how to achieve the oldest form of communication.

But there is another side to unshared rules or mismatched expectations: under some circumstances they may cause pain. The sensory nerves in the skin and muscles and joints are highly specialized. Some specialize in the length of a muscle, others in the movement of just one hair, some in cold temperatures, others in heat, and some are difficult to set off except under conditions likely to cause damage to the cells. These latter sensory types are one kind of "pain" neuron, analogous to a fire alarm tripped by a high temperature. Other sensory neurons respond to light rubbing of the skin but also to severe tissue-damaging stimuli; just how the spinal cord and brain look at the discharge pattern of that neuron and tell the difference between rubbing and tissue-damage is one of the prime challenges of sensory neurophysiology. They probably compare patterns of activity in a number of sensory neurons—take a poll, as it were—before deciding that it is worth labeling as noxious enough to be what we call "pain." What happens if the poll yields a bizarre pattern (say, activity in the "cold" sensors as well as in the "hot" sensors)? Cold and hot simultaneously may not fit any previous schema, may be outside the experience of that individual's central nervous system. Now, under benign circumstances, such a novel input might be intriguing, even a humorous surprise. But under other circumstances, such a novel combination of inputs may elicit feelings of distress, of pain—just as a young bird crouches down in fear at any unfamiliar shape flying overhead (this was once considered to be evidence for an innate "chicken hawk" template)—now it is known to be any unfamiliar bird-sized moving shape overhead. Stimulus configurations that do not match any of the usual templates, which are thus not recognized, can give rise to either extreme of emotions.

This "mismatch" theory of pain, as I am wont to call it, comes to mind when one sees a person in chronic pain from a damaged nerve, who cannot find adequate words (templates) to express the nature of the sensation: it isn't exactly dull aching, or lancinating, or prickling, or warm—rather the patient calls it "weird." And judging from the ways in which damaged nerves spontaneously initiate nerve impulses, it seems likely that they are indeed receiving a barrage from all kinds of sensors simultaneously. Now, under other circumstances, novel patterns of input are funny—the barrage during tickling, for example. The humor of the surprise ending in a joke, to take an example from higher intellectual function. Could we, with some re-education of chronic pain patients, convince them that their disorderly sensations are funny? Or maybe neutral? But at least not painful?

This variegated view of pain research may seem a bit jumbled, but I assure you that it is representative of the state of things—a touch of clinical improvement, of new research ideas, of improved social organization for providing expert treatment and doing interdisciplinary research, and a sad problem of disproportionately small resources for improving things. It is a period of intellectual growth, even if there is little expansion of facilities and personnel. Still, it is frustrating to see all those patients and friends who suffer needlessly because the good ideas still have not been translated into good techniques for dealing with pain.

NEUROLINGUISTICS

When we study human language, we are approaching what some might call the "human essence," the distinctive qualities of mind that are, so far as we know, unique to man.

Noam Chomsky, Language and Mind

No matter how eloquently a dog may bark, he cannot tell you that his parents were poor but honest.

Bertrand Russell

The great difference between a human being and the cleverest chimpanzee is that the human being can have images of persons, times, and places that have not been personally experienced.

Kenneth E. Boulding, Ecodynamics

Social insects behave like the working parts of an immense central nervous system: The termite colony is an enormous brain on millions of legs; the individual termite is a mobile neurone. This would mean that there is such a phenomenon as collective thinking, that goes on whenever sufficient numbers of creatures are sufficiently connected to each other. It would also mean that we humans could do the same trick if we tried, and perhaps we've already done it, over and over again, in the making of language. . . .

Lewis Thomas, in Discover magazine

Much language is symbolic and poets teach us to use words with special force. We may need their help in finding new ways to talk about brains.

J. Z. Young, Programs of the Brain

15

Linguistics and the Brain's Buffer

In the beginning was the word. But by the time the second word was added to it, there was trouble. For with it came syntax. . . .
John Simon, Paradigm's Lost

I don't want to talk grammar. I want to talk like a lady.
Eliza, in George Bernard Shaw's Pygmalion

Strange though it might seem as an introduction to the study of the spoken word, consider the visual world. Color, pattern recognition, neuroanatomy, optics, illusions, contrast, composition—there is no end of it. Each topic is a subspecialty. But the reductionist methods of neurobiology have made headway, analyzing the wiring of the retina, studying the transformations that take place at each relay station in the brain (Chapter 11 of *Inside the Brain* relates that story; David Hubel and Torsten Wiesel won the Nobel Prize in 1981 for their role in it). Many aspects (such as color) are still unsettled, but the neurai templates (usually called "receptive fields") discovered by the reductionist approach now aid everyone's thinking.

If vision seems like a large subject, consider language: speaking and gesturing, listening and watching, phonemes, words, syntax, rhythms and emotional colorations, writing—there is no end to the subject, partly because it is constantly changing, indeed probably still rapidly evolving, unlike the visual system. Just as the natural sciences split off from philosophy several centuries ago, so language's traditional relation to philosophy has begun to be replaced by ties to psychology, neurophysiology, and evolutionary biology. We now think of listening to speech in somewhat the manner of cognitive psychology: that there are a series of templates, or schemata, in the brain to which sound sequences are matched, to which word groups are matched. That the correctness of a sentence (whether it has "good grammar") is a matter of goodness-of-fit to one of the schema formed when you learned the language. Our thinking about language understanding is tending toward the model established by those templates for visual patterns that proved so successful.

This chapter, and the one that follows, will emphasize this template thinking and also another important notion: that of an increasingly large "buffer" memory. Buffers hold on to something temporarily; for example, when you dial a telephone number, the central office machinery has a buffer that holds the numbers until the last one is finished, then it spits them out in one rapid burst to a distant switching computer (you can often hear this as a rapid series of tones when the call is going through). The size of the buffer is often seven digits long, with larger buffers for long-distance dialing. Your brain "buffers" a string of numbers which the Directory Assistance operator tells you until you can dial it. A buffer is a scratch pad, not intended to save something forever but to hang on to it long enough for some operation to be carried out.

For some buffers (like a grocery list), order is unimportant. But for the buffers we will be considering, keeping things in the right order is terribly important (jumbled phone numbers are worthless, jumbled sentences mean something different), and we will be concerned with buffer size for much the same reasons that elementary textbooks use shorter sentences than advanced ones: more complicated propositions require longer word buffers in the brain.

From the outside, language is studied by us all, experts that we are in the acquisition of new words, in effortlessly translating a series of sounds into a meaning and acting upon it. Children are especially attuned to acquiring new words by imitation (it was noted over two centuries ago that European children had no trouble learning the Hottentot language, even if their parents couldn't); just like parrots, the children may try to produce the sounds they hear. But unlike parrots, they also invent new words and try them out: if they learn "bad" and "dad" they may try out "bab" and "dab." Infants seem to be born with an ability to distinguish many phonemes; e.g., to tell /ba/ from /pa/, where the major difference is the repeated vibration of the vocal cords near the beginning of /b/.

Soon children speaking words are trying out new ways of combining them. They may say "where it is?" instead of "where is it?" and only gradually realize that others do not use this arrangement of subject, object and verb (except, of course, in saying "Show me where it is"). We build up a series of acceptable templates or schemata for phoneme order (words) and for word order (phrases) which allow us to make sense out of a sentence we have never encountered before, such as this one.

Linguistics did not start from this neurobiological perspective but from the comparative study of languages. The common features of all languages led to the notion that the regularities between otherwise dissimilar languages might have something to do with biology, with features of the brain.

The deep structure of language is defined, more or less, as that which allows us to say that two sentences are synonymous; e.g., "Mary washed John" is the same relation of actor-action-object as is "John was washed by Mary." Linguists argue about just how deep the deep structure should be removed from the surface structure of the sentences, but many of their difficulties are related to trying to create a one-way series of processing steps—the little boxes with labels connected with arrows—relating syntax (word order) to semantics (meaning) and to surface structure (grammar) and phonetics (e.g., inflections).

Naturally, linguists have attempted to infer a series of steps in the production of a sentence. But there seem to be too many feedback loops, as when the meaning of the sentence (deep structure, presumably) is changed at the last moment simply by the voice rising or falling at the end of "You are happy." Which, of course can communicate a lot: Maureen Howard tells of growing up in a neighborhood "where by a mere inflection of 'Hello' you could tell that the pert young widow had lost her beau, the Montours were coming up in the world this week, the Drews had not paid their grocery bill." Indeed, from the single word "Hello" at the beginning of a phone call, you can frequently identify a unique individual in the whole world; even if you don't recognize the individual, you can still make a good estimate of the speaker's sex, age, friendliness, nationality, education or socioeconomic status, and health. If one word can convey an enormous amount of information far beyond its literal meaning, then it is hardly surprising that linguistics has trouble with simple theories for the production of word order.

That deep structure is innate and preprogrammed is perhaps the best known of Noam Chomsky's arguments, made from the comparative study of languages and similar sources. This notion seems to have come as a revelation in those areas of the humanities whose neurological vocabulary is limited to *tabula rasa,* though I have yet to meet a biologist who was surprised. Linguistics is, for better or worse, common ground for the humanities and the natural sciences, one of the major places where C. P. Snow's two cultures meet. And deep structure is a major point of congruence.

Though seldom expressed in neurological terms by its enthusiasts, innate deep structure is just another way of saying that we must seek a neurological basis of the linguistic actor-action-object equivalence relations in the anatomy and physiology of the brain. If language is a system which is superimposed upon another older sensory-motor-regulatory neural system, then we might hope to see the actor-action-object paradigm in the more primitive system.

In the throwing madonna painting, one sees actor-action-object: it depicts a mother-with-infant, a throwing motion with rock in flight, and the rabbit target. Most motions outside of simple locomotion have goals. Ballistic motions, in particular, are simply characterized by a starting point and an end point with a rapid motion in between. The actor is self: self, motion, goal. In looking around the environment, the actor may be another subject of interest: a deer approaching a salt lick. Like the difference between scalar and vector qualities in mathematics, the representation is not a simple static set of interrelationships: motion, or at least a direction, is suggested. And so one wonders whether actor-action-object is reflected in sentence structure in more than a casual way, whether subject-verb-object is an offshoot of a nonverbal paradigm.

Spoken language output is sequential (sign language may, in contrast, express various sentence elements simultaneously). Furthermore, word order plays an important role ("John washed Mary" has a quite different meaning from "Mary washed John"). Tactical considerations in language also influence the choice of the active form "Mary washed John" over the softer passive construction "John was washed by Mary"—or the invention of an even more pronounced circumlocution.

This emphasis on sequence tends to bias our view of language processes, as witness the sequential black boxes that give linguistic theorists such difficulties. But the premotor aspects of spoken language need not be any more sequential than the premotor aspects of movement planning: in deciding when and where to pounce, my pet cat is simultaneously taking into account all sorts of factors including her present posture, the bird's position, and strategic considerations such as whether the bird is too big to tackle (or whether I am watching). The fact that the pounce starts with extension of knee and hip joints does not mean that this aspect was planned first or last.

The subject-verb-object (SVO) order of most English language sentences ("Borg served an ace") happens to correspond with the stated order of the actor-action-object paradigm. Is that important? The Japanese might consider it a bit of cultural imperialism because typical Japanese sentences have the verb last (SOV: "Borg an ace served"). At least in stating complicated logical propositions, there are alleged to be some advantages to the SOV order. The arguments can be heard among students having the two different kinds of hand calculators: those that use the traditional "A plus B equals" keystrokes mimicking SVO and those requiring the SOV-like reverse Polish format "A, B, plus." Or among the computer enthusiasts similarly favoring the SOV-like language called FORTH. Essentially, one can avoid clustering ambiguities—which otherwise require the liberal use of parentheses and brackets—in reverse Polish.

In German, the verb is last only in dependent clauses: "If SOV, then SVO." In classical Arabic, the verb comes first: VSO ("Served Borg an ace"). Indeed, of the six possible permutations of the three items, all six are used as the standard in one language or another (it was once thought that OVS was an exception, but Hixkaryana is now known to use it: "An ace served Borg"). A language's choice of major word order does seem to often have implications for the minor words: for example, prepositions in Japanese are postpositions ("by bus" would come out as "bus by").

All of which would indicate that, while sentences are being assembled in the brain's buffer, there is no mandatory biological order, only that dictated by local convention and long practice. It is probably as reasonable to say that deep structure reflects actor-object-action as actor-action-object. Alas, the throwing order does not help us understand word order, at least at this level.

It may, however, be worth pursuing word-order cues when the language is not overlearned and has a different word-order preference than the native language. I can remember being distressed by SOV when learning German: to have to read through to the last word in the clause before the action was revealed seemed like an interminable wait (and at my reading speed, it was). It led Mark Twain to comment in *A Connecticut Yankee in King Arthur's Court*: "Whenever the literary German dives into a sentence, that is the last you are going to see of him till he emerges on the other side of the Atlantic with his verb in his mouth."

One can, of course, speculate endlessly on the philosophical implications of such word-order preferences for national personality types: that SOV might produce, for example, a more contemplative personality because one must suspend judgment until all the facts are in—rather than, under English-style SVO, impetuously getting the action in motion before its target is revealed.

The brain's buffer for words is also essential when trying to fathom a long sentence with many clauses. You need to analyze each subject-verb-object group, translate it into deep structure of actor-action-object, repeat the analysis for each phrase of the sentence, and then relate them to the sentence taken as a whole, performing yet another actor-action-object deep structure job on the overall reduced sentence. (The in-group term for repeating the action over and over for improving results is "recursive," as in that tracing-over exercise from first grade when you learn cursive handwriting.) The rules of word order (grammar) have to be applied recursively, and to do that you need a big sequential buffer to store the whole sentence temporarily.

But even 5-year-old children can accomplish that. Consider the nursery rhyme about the house that Jack built:

> This is the farmer sowing the corn,
> That kept the cock that crowed in the morn,
> That waked the priest all shaven and shorn,
> That married the man all tattered and torn,
> That kissed the maiden all forlorn,
> That milked the cow with the crumpled horn,
> That tossed the dog
> That worried the cat
> That killed the rat
> That ate the malt
> That lay in the house that Jack built.

Each phrase is analyzed and then substituted for the "That" of the next line, simply becoming the new actor. But adult sentences have embedded phrases that can modify each of the actor-action-object triumvirate: "I believe Jack says you think you heard him."

You need a good holding buffer, and as noted in Chapter 5, evolution has long had an increasing need for short- to medium-term memory storage as social life has become more complex, as omnivore tendencies to try out new foods have expanded. But an extremely sequential memory, rather than a general gestalt memory. What could have selected for a sequential buffer? Ballistic motor sequencing is one candidate.

The motor buffer is perhaps more easily understand when it is used for speaking a sentence. If language was planned using a motor buffer for storing the sentence, many factors could operate simultaneously and interactively upon that sentence buffer: semantics, phonetics, and transformation rules. One could be speaking while planning ahead, with the words already spoken serving as a constraint on subsequent syntax; as the Roman poet Horace said, "Once a word has been allowed to escape, it cannot be recalled." But otherwise, there might remain much freedom to create alternative sentences which all satisfy the same desired actor-action-object interrelationships. As when we change our minds in mid-sentence and complete the sentence with a new ending, but fitting it to the grammar of what had come before.

But a motor buffer for a sensory decoding task? Strange but probably true: The learning-by-imitation theory has been extended to listening-using-(subaudible) speech to account for various phonological and physiological evidence. The motor theory of language perception and the cortical mapping evidence (discussed in chapters 4 and 16) say that we are using some motor templates to sort phonemes into categories (telling the difference between /ba/ and /pa/ by using the premotor program for forming those phonemes), though with the muscular output suppressed at some premotor stage.

Two aspects of motor systems have entered our discussion of language: the concept of motor templates in a sensory perception task, and the concept of a motor sequencer—in particular, a buffer into which proposed actions can be stored during "get set" and then rapidly emptied at "go."

Yet the same buffer could hold the incoming string of phonemes, allow schemata to identify the words, then allow surface and deep structure to be analyzed—recursively, if there are embedded phrases. In other words, deep structure could communicate simultaneously with the other levels of both language reception and production.

Though the examples given in this book tend to imply that the buffer was shaped by throwing success and that the sequential actor-action-object is a paradigm derived from throwing stones, this need not be the only deep paradigm on which language could operate. The equivalence relationship, used by the cat to know that creeping up on the bird from behind will accomplish the same actor-action-object relationship as a frontal attack, is also what we use to tell that two sentences are synonymous or that *sui generis* means about the same as unique or peculiar.

Which brings up the contention of some linguists that language is a human capacity *sui generis,* without parallel in the animal kingdom or in other human systems. Though probably not intended in the same spirit, this always reminds me of the creationists' argument for special creation (because humans are so special, they must have followed a different path of creation than mere animals). We may yet come to conclude that human verbal language is unique, but first we must try out some more mundane hypotheses. Such as evolution from gestures. Such as evolution from species-specific vocalizations. Such as the possibility that language was built atop a motor sequencing system which greatly enlarged for another "reason" (throwing range and bigger prey) and selection pressure (hunting success, as opposed to communication success). Such as the idea that recognition may consist of fitting incoming information to templates or schemata, just as syntax may correspond to how well an utterance fits with a schema of prototypical word-type sequence formed by an experience-modified innate premotor program.

So sequence and grammar are all important—but, alas, I still cannot diagram a sentence. That is likely left over from high school. I was comparing notes with a classmate of mine who got D's in English but still managed to become an editor of the *Los Angeles Times* (he still refuses to diagram sentences). We couldn't stand the diagramming exercise because we already knew what was correct from long experience with language. Taking a sentence apart into pieces seemed pointless, since we already knew what sounded correct and what didn't. It wasn't until I tried to learn German in college that I suddenly realized what grammar was all about: spotting the different parts of the sentence is quite essential if your vocabulary is shaky.

I now suspect, having been further exposed to examples of overquantification and "physics envy" in the softer sciences, that our teacher was trying to make English grammar into a rigorous subject comparable to Euclidian geometry. But language isn't like that: it is naturally a bit loose and always evolving (just try reading something written 200 years ago in the original).

There are some foundations underneath, as Chomsky noted, and they may rest on some nonlanguage specializations of the brain for another sequential task needing a big holding buffer.

"Get set" and "throw" as a foundation stone for language? Stranger things than that have happened in evolution. Birds, for example.

16

Probing Language Cortex: The Second Wave

The highest activities of consciousness have their origins in physical occurrences of the brain, just as the loveliest melodies are not too sublime to be expressed by notes.

W. Somerset Maugham

The problems and findings at other levels are usually largely irrelevant for those working at a given hierarchical level. For a full understanding of living phenomena every level must be studied but . . . the findings made at lower levels usually add very little toward solving the problems posed at the higher levels. When a well-known Nobel laureate in biochemistry said, "There is only one biology, and it is molecular biology," he simply revealed his ignorance and lack of understanding of biology. *Ernst Mayr*

Working from the bottom up, neurobiology focuses upon ion channels, collections of channels and sheet insulation called membranes, diverse collections of membranes called nerve cells, collections of varied nerve cells called circuits, modular collections of circuits exemplified by the leech's ganglia and the visual cortex's hypercolumns—but then what next?

Working from the top down, we can distinguish linguistics, instinct, cognition, memory, hemispheric organization, cortical maps, but then . . . ? How do modular super-hypercolumns generate grammar, store the word "rabbit," recognize a fuzzy animal as a rabbit, pronounce the word "rabbit"—or set about throwing a stone at a rabbit? What are the missing levels that link our bottom-up and top-down approaches to the brain? Some idea can be gained by taking the language system of the brain and making some guesses about the missing links.

Not only is language a system of prime importance in our endeavor to understand ourselves, but it has visual aspects (reading), auditory aspects, memory aspects (as in naming, or recalling a recent word), emotional and selective-attention aspects, advance planning aspects, and motor aspects. Even a partial understanding of how it integrates these aspects could prove most helpful in approaching the frontal lobe role in strategy, the right temporal lobe role in emotional face recognition, or the parietal lobe role in keeping track of "extrapersonal space" into which one might move a limb.

The story so far is still monolithic enough to be told as a historical narrative. A century ago, there was much emphasis upon the size of the head, many attempts to find a bigger-is-better correlation (recounted recently by Stephen Jay Gould in *The Mismeasure of Man*). Since the cerebral cortex was obviously the part of the brain that was so much bigger in humans than in the brains of the great apes, it was considered the seat of higher functions such as language. The French neurologist Paul Broca had even identified some parts of the cortex in which stroke damage caused aphasia, and the identification of language with cortex was considered firm.

Language physiology, as opposed to the evidence from aphasia, started in the 1930s; Otfrid Foerster in Germany was a pioneer, but the most extensive observations, spanning three decades, were made in Montreal on epileptic patients undergoing brain operations under local anesthesia. Wilder Penfield, a neurosurgeon with training in physiology, used electrical currents to stimulate the surface of the brain as part of making a brain map for that particular patient, prior to deciding exactly which epileptic parts of the brain could be safely removed without causing more trouble than the epilepsy he was trying to cure. This technique was developed by the pioneering Boston neurosurgeon Harvey Cushing in 1909; what Foerster and Penfield did was to apply it to the study of language.

Penfield's researches, especially those done in collaboration with Herbert Jasper and Lamar Roberts, revolutionized our concept of brain maps in the 1950s. They showed aspects of both language and long-term memory. There was a region in the frontal lobe, corresponding to the "Broca's area" of the aphasia literature, where stimulation blocked language output: the right words would come out just as soon as the stimulation stopped. There were sites where the patient could talk but not find the right word (anomia: "That's a . . . , you know, the thing you put in a lock"). And there were sites where specific memories could be evoked: the so-called experiential responses. These "evoked memories" were sometimes musical when they occurred (which is in only a small percentage of patients tested). They always make for a good story: one young Seattle man heard rock music every time a certain right temporal lobe site was stimulated. And he was sure that it was music by a particular rock group, the Led Zeppelin. The Montreal patients seemed to have more classical tastes. But in any case, it is a rare phenomenon and probably part of the patient's epileptic process rather than a normal memory process.

But this early work in Montreal was essentially the first look at *terra incognita,* and the more detailed exploration and mapping did not occur until a quarter-century later. This "second wave" took place primarily in Seattle, and while it grew out of the Montreal tradition, it also retraced a *fin de siècle* argument of the Paris school of neurology and linguistics.

Language in the Depths of the Brain

A French neurologist, Pierre Marie, said that the deeper (and hence more primitive?) parts of the brain such as the thalamus also played a role in language. This caused another stressful argument in the Parisian academic circles about the turn of the century, reminiscent of the arguments that had caused the Linguistic Society of Paris to ban any presentations on the origins of language starting in 1886 (they reconfirmed the ban in 1911). There the issue lay, for six decades.

In the early 1960s, many neurosurgical centers were treating Parkinson's disease by controlled damage to part of the thalamus in the brain's depths, essentially trying to bring the muscle stiffness system back into balance again. Arthur A. Ward, Jr., was building up a physiologically oriented neurosurgery department in Seattle at the University of Washington; he had trained in one of the leading physiology departments of the day and then taken his neurosurgical training with Wilder Penfield. One of Ward's specialties had become the treatment of Parkinsonism by sticking a long needle into the thalamus under x-ray control, identifying the sites physiologically, and then heating a small area to destroy it. At the time, he was training a young neurosurgeon, George A. Ojemann. They noted that when they electrically stimulated the thalamus at subliminal levels prior to making the destructive lesion, the patients had trouble talking. Language in the thalamus? Shades of Pierre Marie.

Being good physiologists, they began to plan how to take this surgical opportunity in awake patients, in order to learn more about what language was doing in the thalamus. George Ojemann went off to the Clinical Center at the National Institutes of Health near Washington, D.C. Arriving at the same time was a psychology Ph.D., Paul Fedio, and together they started up a project to test for language in the thalamus. They designed a series of tasks for the patients to perform that would test specific aspects of both language and memory for verbal material: the patients would watch a slide show on a movie screen in the operating room and name the objects they saw, then recall them from memory some seconds later after performing another task designed to keep them from rehearsing (counting backward by three's is such a consuming mental task). They were originally interested in naming: if the patient didn't respond during stimulation, was consciousness being altered? But as we shall see later, it was the memory aspect that proved to be the most interesting.

Two years later, Ojemann returned to Seattle to join the faculty and started thalamic experiments in earnest. Ward had designed a neurosurgical operating room (O.R.) ideal for physiological experiments, with copper shielding built into the walls, ceiling, and floor to prevent electrical interference (remember the screen wire cage in Jerusalem?). This was not as essential for the thalamic operations as for the epilepsy operations, as Ward had continued the Montreal tradition of doing epilepsy operations under local anesthesia, using physiological mapping of the exposed cerebral cortex to help make decisions and to further our basic knowledge of the human cortex. It was the combination of the two surgical procedures typically done under local anesthetic—thalamotomy on Parkinson's victims and cortical removal for epileptics—that proved to be so important in the evolution of the language and memory studies, in what became the "second wave" of language physiology.

Language Localization in the Cerebral Cortex

The slide show for testing Parkinsonism patients in the O.R. was gradually used during epilepsy operations as well. Neurosurgeons have often tested for language during epilepsy operations: the location of language areas cannot be judged from a standard map, as humans have proven to be quite variable, with many individual aspects to where language is located. But the testing was often informal, making it hard to progress past the point attained by Penfield and Roberts. With the slide show and stimulation protocols developed for the thalamic operations, a new era opened up in cortical language physiology. At the NIH, Paul Fedio and John Van Buren made a series of observations, beginning just before George Ojemann left to return to Seattle and continuing until about 1970. In Seattle, Ojemann tested his own patients as well as those operated upon by fellow neurosurgeons. But while the thalamic observations were frequent, progress was slow on testing cortical language organization, with about one patient per year tested up until about 1976, when the pace quickened. These observations established that there were sites in the cortex that disrupted short-term memory for verbal material (stimulation while the patient was counting backward may not have affected the counting, but it certainly affected later recall of the prior material).

Several things happened in the mid-1970s to speed the language research. First, there was the completion of another research project being done on epileptics in the O.R., where Ojemann, Ward, and I studied single neurons in the epileptic brain, comparing their firing patterns to those of induced epilepsy in research animals, trying to keep the animal research on the right track by comparing it to the real thing. Its completion meant that some time, always exceedingly valuable in the O.R., was freed up. But most important was a sabbatical year spent in Seattle by a neurolinguist, Harry Whitaker, editor of the journal *Brain and Language* and now a professor at the University of Maryland.

"Whit" had for many years studied patients with various injuries to the brain (his studies with Maureen Dennis, of language moving from the left to the right brain in children with congenital left-brain damage removed by surgery, were mentioned in Chapter 10). And here was the opportunity to apply many neurolinguistic techniques to fairly normal patients in which the stimulation produced only a temporary blockage—harmless and on demand, at the press of a button on an electrical instrument. The neurolinguists' tricks-of-the-trade began to be adapted to the neurosurgical and physiological setting.

The Bilingual Brain

Ojemann and Whitaker found that the standard maps of language were, in fact, only average: language maps were indeed quite individualized, just as are our faces. One of the particularly memorable studies initiated during Whit's sabbatical resulted in an article entitled "The Bilingual Brain," where bilingual epileptics were tested in both languages. Electrical stimulation of some sites affected performance in naming simple objects in both languages—but at other points, it would affect only English and not the other language, or only the other language and not English. Just moving 5 millimeters along the top of a gyrus could change the language that was disrupted.

Separate brain for different languages? Bilingual patients are not often seen in the U.S.; the definitive study on the subject will probably be done somewhere else in the world where people are typically multilingual. It would be especially desirable to compare closely related languages and widely different languages (the languages compared to English in Seattle happened to be Dutch, Spanish, and Greek—all from the Indo-European family of languages). A start has been made on a more diverse language comparison. Richard Rapport did his neurosurgical training in Seattle and then spent a year in Kuala Lumpur, where the three Malaysian epileptics on whom he operated were fluent in both Chinese and English. They too had separate sites for each language, plus the usual overlapping sites.

Sex and IQ

The individual variation from an average map is not just a random business; some of it is clearly related to sex and IQ, as Catherine Mateer, George Ojemann, and Samuel Polen (a professor of speech pathology and audiology on sabbatical from Western Washington University) have recently shown. The area of cortex where naming can be disrupted is somewhat smaller in women than in men. This had been suspected because left-brain strokes cause aphasia much less often in women than in men.

The maps also vary with the verbal IQ of the patient (in a manner independent of sex). Only one of nine patients with an above-average verbal IQ had naming disruption from the parietal lobe. Seven of nine patients with below-average verbal IQs (range 69–96) had naming disrupted at parietal lobe sites. The issue here is not the total amount of cortex devoted to language, but its extension into the parietal lobe above the Sylvian fissure. And the above-average-IQ patients might also use that cortex for language, with stimulation there simply incapable of disrupting the rest of a well-organized system elsewhere; it simply isn't known how to interpret the results yet. It seems likely that some arrangements of language cortex are more efficient than others.

Is Language Superimposed on Motor Sequencing?

Another important development took place when Catherine Mateer came to Seattle in 1977 as a postdoctoral fellow (she later joined the faculty). As recounted in Chapter 4, she and Doreen Kimura had established that left-brain strokes affecting language also disrupted motor sequencing, for both hands and for both sides of the face. Mateer and Ojemann then went on to test oral-facial sequencing in epileptic patients in the operating room. As expected, stimulation disrupted sequences but not individual actions. The big surprise was the 86 percent overlap in the sites where this occurred and sites where phoneme discrimination was disrupted. A purely motor task and a purely sensory task, both living at the same sites? Did imitation lie at the heart of recognition?

But actually this intimate relationship had been predicted by the linguists who noted that the categories for phoneme discrimination seemed to parallel the difficulty of pronouncing them. This "motor theory of speech perception," established in an influential review in 1967 by A. Liberman and co-workers, suggested that we detect a speech sound (a phoneme) by comparing it to a motor schema or template for pronouncing it. Just as some people move their lips while reading, this theory suggests that we all subliminally move our lips while listening to someone speak. This intriguing interaction between the sensory side of language and the motor sequencing of the mouth has led to many new ideas about how language got started in the cortex surrounding the Sylvian fissure. Kimura was perhaps the first to suggest that a lateralization to the left brain of motor sequencing would provide an important alternative to the traditional theories of language building atop gestures or other communication circuits.

Verbal Memory Surrounding the Central Core

Just as for thalamic memory testing, the epileptic patient is shown a slide depicting a common object and responds, "This is a dog [or ball or telephone].” A second slide gives the patient another task to do for distraction, and then a third slide prompts the patient to recall the object on the first slide: “Dog.” If stimulation during the second-slide distraction task causes errors on third-slide recall attempts, the site of stimulation becomes known as a memory site (under certain conditions, first- or third-slide stimulation can also produce discrete memory-type errors). A map of such sites varies quite a lot from patient to patient, but they typically form a rim around a central core of sites where phoneme discrimination and oral-facial motor sequencing are disruptable.

Recall errors can also occur from stimulation during the first or third slides. Sites causing third-slide recall errors from first- or second-slide stimulation (“input” and “storage” phases) tend to cluster in temporal and parietal lobes; sites causing recall errors during third-slide stimulation (“output” phase) are more common in frontal and parietal lobes.

Semantic Sites and Syntax

There are some sites, typically located near the interface between core and memory rim, at which stimulation disrupts naming on the first slide. The patient will say, “This is a . . . you know, it’s a . . . thing you put a key in.”

There are sites at which only reading is disrupted. Given a future-tense sentence to read with a blank near the end of the sentence to fill in (“If my son is late for class again, he’ll ______ the principal”), the patient substitutes jargon, or repeats himself, or may be unable to complete the sentence. (For example, a reading error is the response “My son will getting late today he’ll see the principal,” where the noun and verb stems are correct.)

Still more interesting are the sites where the only known effect is to disrupt the good grammar of a sentence. In a sentence completion task like the one above, the patient will read it correctly, but in filling in the blank will make minor errors in tense or word order, though not enough to destroy the meaning of the sentence (see Chapter 4 of *Inside the Brain* for examples).

Ojemann sees the overall maps as suggesting a central core for motor mimicry and phoneme discrimination (in frontal, parietal, and temporal lobe gyri near the Sylvian fissure; thus "periSylvian"), a surrounding rim of regions specializing in short-term verbal memory (though with recall aspects more frontal, and input and storage aspects more temporal), and at the margins between these two major regions, a patchwork of specialized sites related to syntax, naming, and reading. Looking back at the extensive literature on aphasia, it is apparent that in most cases, part of the central core has been permanently damaged. Damage to other regions may cause only temporary deficits.

But this patchwork-quilt picture of language cortex says little about how the individual areas work together to produce language, to recognize and remember words. Language is surely a committee effort of these areas. At some point, our reductionistic approach must reverse, to incorporate the broader picture of interacting regions—just as a portrait of a legislative body cannot be merely a series of chapters on the individual members but must analyze the shifting coalitions that group the members together for some purposes but not others, and must analyze the emergence of leadership.

Selective Shifts of Attention

The most interesting outcome of the thalamic language studies that started the second wave was some memory effects. Stimulation during the first slide might not affect correctly saying "This is a ball," but it would increase the chances that "ball" would be the reply on the third-slide recall attempt after distraction. Stimulation during the third-slide recall attempt didn't always cause recall errors, but it increased the chances of an error beyond that expected without stimulation. Stimulating during both first and third slides might cause no change in the unstimulated error rate, as if the effects averaged out.

The simple interpretation of this result favored by Ojemann is the following: Stimulation taps into the brain's specific alerting system (a subdivision of the reticular formation's sleep-wakefulness system). Stimulation orients the patient to incoming information from the visual world (indeed not just any information; left-thalamic stimulation affects only information with verbal significance). Thus stimulation during recall attempts shifts attention away from internalized information in short-term verbal memory, making errors increase. Stimulation during the original presentation causes the patient to pay even more attention to what is on the first slide, thus increasing chances that the information will be recorded vividly enough to be recalled after the distraction.

The thalamus contains many pathways, not only to the reticular activating system, but connecting to the cerebral cortex. Such as language cortex. Might the thalamus be acting somewhat like an orchestra conductor, tuning down the activities of some of the patchwork regions of language cortex while enhancing others, then changing things around as the tasks shift? Does the thalamic story fit with the cortical story?

To tell the story requires another traveling-neurobiologist tale. In the aftermath of the 1973 Yom Kippur war, a young Israeli, Itzhak Fried, went to Tel Aviv University and studied physics. He decided to do graduate work in psychology and was admitted to the doctoral program at UCLA. UCLA has long been a center for the study of the reticular activating system. Fried not only learned to study the minute voltages seen on the scalp from brain activities preceding a response ("event-related potentials"), but he got mixed up with the neuroanatomists studying human brains, trying to spot anatomical differences between the left language area and the corresponding points on the right side of the brain. One of his mentors, Arnold Scheibel, a professor of anatomy and psychiatry, was familiar with the work being done in Seattle on language cortex, and so George Ojemann came to UCLA in 1978 to lecture on how variable the naming sites were between patients. Fried promptly arranged to come to Seattle and work with Ojemann for part of his Ph.D. dissertation in psychology at UCLA. (If this sounds roundabout, be assured that it is almost traditional in neurobiology—many of us have such a story.)

Besides stimulating, one can electrically record directly from the exposed cortical surface. This is routinely done with an EEG machine for helping to define the epileptic region. But it can also be used to examine a site's responses to a slide being flashed on the screen. The slides were simple black-and-white sketches of common objects, as usual, but with a bold slash across the whole picture at one oblique angle or another.

When a region of brain is just idling, the EEG waves are larger and slower than when it is particularly active (we say that the EEG is "activated" when its bumps become smaller and more frequent, it is rather like how an automobile engine smooths out when accelerated from a rough idle). When the patient was asked to name the object shown on the slide (silently, to avoid complications from muscle commands), certain areas of brain would be activated. This was not general but in a patchwork; when stimulation mapping was done later in the operation, such EEG activation sites demonstrated anomia ("You know, it's a . . . a . . ."), whereas adjacent sites where activation did not occur did not exhibit stimulation-evoked language changes.

Furthermore, this selective EEG activation occurred only if the patient was paying attention to the verbally codeable aspects of the picture; if the patient was asked to pay attention to the angle of the slash across the picture of the object, the left-brain sites would not be activated (presumably right-brain sites were, but one cannot record from both sides of the cortex at the same time in the same patient). The same slide produced different responses, depending upon what the patient was set to do.

There are many aspects of the nervous system over which we have little voluntary control, such as autonomic regulation of heart rate and blood pressure. But selective attention for verbal versus pictorial aspects of our visual world would seem at the opposite end of the voluntary-involuntary spectrum, near the heart of higher brain functions. Here we have seen such selective attention mechanisms at work in language cortex. Self—what makes you and me different—is surely all tied up with how we use such selective attention mechanisms to engage ourselves with different aspects of this multifaceted world.

Is the Temporal Lobe Really Temporal?

It is obvious why the frontal lobes were named "frontal," but the etymology of "temporal lobe" is less apparent. Is it because the temporal lobe has a clock? Perhaps, but that's not why it was named that.

The temporal lobe was named after the temple: the spot we often rub between eye and ear has the tip of the temporal lobe just behind it. But Robert Efron tells one that the temple was named for time (via the Latin *tempus*) in an amusing sense: as we age gracefully, the temple is the first place that the gray hairs appear. So the temporal lobe is, in a round-about way, named for time. If that piece of brain has something to do with time, however, it probably operates on a much shorter time scale, milliseconds rather than decades.

The function of the tip of the temporal lobe has always been a bit of an enigma. The reason that neurosurgeons can remove it to treat seizures is because so little happens when it is missing. If the removal does not damage the language sites farther back in the temporal lobe, the deficits are quite subtle. Thus our usual method for assigning function (deficits from strokes and similar damage) fails us entirely. Even stimulation mapping usually fails to show functions there, at least with the tests used thus far. Neurosurgeons have been known to suggest that its function is to cause temporal lobe epilepsy, as that is what changes when the temporal tip is removed.

This has suggested that the temporal tip forms some sort of redundant system. A passenger plane has several backup systems for lowering the flaps in case the primary system fails; remove one and you'd never notice it was gone unless the two other systems failed first. There are two special sets of data that indicate that the temporal tip has the capacity to function as part of language's motor system.

A patient with a slowly growing tumor near the traditional Broca's area of the left frontal lobe was mapped; the usual response of Broca's area to stimulation is arrest of all speech output (the patient tries to talk but cannot even say "This is a . . ."), but stimulation had no such effects in this patient's frontal lobe. Slow damage can cause rearrangements even in adults, where another area takes over the function, and since the patient was only mildly aphasic, this seemed likely. In this patient, Broca's area responses were found below the Sylvian fissure, 3 centimeters from the temporal tip, showing that this enigmatic region has sufficient motor system connections so that it can, if necessary, function as the motor output specialization of the language system. In the second special patient, only finger spelling was disrupted at the temporal tip; this woman could hear but had learned to finger spell as part of her work with the rehabilitation of the deaf. This again shows that the temporal tip can function as part of a motor sequencing system.

Careful testing also shows that temporal tip removal does cause some changes in performance on an auditory judgment. Suppose that you listen to a pair of tones, one high and the other low—except that the order is sometimes high-low, sometimes low-high, and the patient has to judge which. When the tones are spaced far apart ("Beep . . . . Boop"), the task isn't too hard. But if the second follows close after the first ("Beep-Boop"), the task gets hard. If the tones are presented in the ear opposite to where the temporal lobectomy occurred, the judgment is more difficult than usual and they have to be more widely separated ("Boop . . Beep") than usual before the judgment becomes reliable. If you think this is complicated, I am actually oversimplifying the results—but I did say that the deficits were subtle; it is amazing that Ira Sherwin and Robert Efron discovered the deficit at all. There had, however, been clues that the region had a role in various fine timing judgments; Paula Tallal and co-workers had suggested that the left brain processed rapidly changing auditory signals (of which speech is but one example) to a greater extent than the right brain.

So damage to the temporal tip suggests that the region has some redundant function and that it might be associated with timing. If you read the literature on heart cells beating away in culture dishes, or on models for circadian rhythms in sand fleas, it all starts to make sense. As mentioned in Chapter 4, redundancy allows for more precise timing: if one uses four times as many identical timing circuits in parallel, a cell which averages their outputs will double its accuracy (i.e., its standard deviation will halve). So loss of temporal tip might merely degrade some fine timing capabilities to merely good.

Readers of Chapter 4 perhaps will also remember an evolutionary argument for why precision timing circuits might be exposed to natural selection pressures: If redundant timing circuits can be used for motor sequencing as well as auditory processing (not unlikely, considering the evidence for the motor theory of speech perception), then they could help time the release of the rock during throwing. To double the throwing distance for a rabbit-size target requires an eight-fold narrowing of the launch window. Timing redundancy circuits would allow that to be done with a sixty-four-fold increase in timing circuits applied to the problem. A hominid brain that could temporarily borrow timing circuits from elsewhere, or had more to start with because of brain enlargement, would be a better brain for controlling throwing.

The whole periSylvian region probably is involved with timing and sequencing, as telling one phoneme from a closely related one is a matter of making fine timing judgments on a sequence of tones. Commanding the pronunciation of a phoneme sequence presents exactly the same problem of fine timing within a sequence. The temporal tip may merely be the cortical annex for supplementing the timer array for the really precise jobs.

Redundant timing neural circuits provide an attractive theory for relating motor sequencing and phoneme discrimination, for tying together throwing and language. It even suggests why bigger brains might be better.

"Natural" Maps vs. the Time Window

Someone could do a fascinating doctoral dissertation on the cross-fertilization roles of the traveling scholar in neurobiology. The itinerant graduate student, the postdoctoral fellowship, and the faculty sabbatical have opened up many new avenues, permitted expertise in one area to be brought to bear upon new questions, and prevented many cases of parochialism and wasted effort by stirring the pot.

But language physiology is still a young subject and it should perhaps be emphasized that, though their backgrounds have varied widely, just over a dozen people have had major involvement in it so far, even counting the first wave from Montreal. Contrast this to the traditional source of brain maps, the aphasia from stroke victims. This anatomic pathology version of language investigation has had contributions from thousands of investigators, whose backgrounds ranged the spectrum from the pre-psychoanalysis Sigmund Freud to the post-physics William Calvin. But most of these investigators have been "observers" rather than "experimenters," because nature's accidents perform a messy experiment and we merely document the outcome and try to make some sense out of similar accumulated evidence. The first two waves of language physiology have given a glimpse of a detailed landscape which the stroke documenters only occasionally suspected.

Our existing physiological maps are, however, inevitably biased by the backgrounds of the lucky investigators. As the tests used bear the stamp of their particular scientific upbringing, so they bias the maps made from the results of those tests (consider what might have happened to the maps if their backgrounds had emphasized Freudian concepts rather than Sherringtonian physiology!). Neurobiology is still a small enough world that we can trace genealogies, of who was influenced by whom. One of the auditory tests used in the O.R. was developed by C. E. Seashore at the University of Iowa, where George Ojemann grew up as the son of a psychology professor; later, he happened to acquire Seashore's desk and studied at it all during his time in medical school, only to later discover an interesting use for the Seashore test. The emphasis upon motor sequencing grew out of Doreen Kimura's neuropsychological interests via Catherine Mateer's involvement in test design. My theoretical emphasis upon precision timing buffers, which I have superimposed on the Kimura-Mateer-Ojemann notion of a primary specialization for motor sequencing, grew out of my doctoral dissertation with C. F. Stevens fifteen years earlier on why neurons are noisy. Interesting facts from the past are always popping up, so teachers never know where their influence reaches.

As more people get the opportunity to study language physiology, the emphasis will inevitably change as different backgrounds cause new methods to be brought to bear, new maps made. The answers you get depend upon the questions you ask, the methods you have—and what's actually there in the brain. We must constantly ask whether our classifications are as natural as possible, and not overly biased by our approach.

This second wave of human language physiology has been extraordinarily exciting. The mosaic of functions demonstrated by the present stimulation maps show some of the natural physiologic subdivisions of language with a clarity difficult to perceive in linguistic studies or in stroke impairments on a naturally variable population. As the functional relationships between such areas are further investigated, along with their ontogeny and functional plasticity, additional windows upon the subspecialties of language cortex will be gained. We may expect our definition of functional roles to become even more natural than those growing out of the present youth of language physiology.

Unlike other areas of science, however, this one could abruptly freeze in place. We must hurry, because our opportunities for access to the working human language cortex could disappear. Skulls are opened and language cortex mapped with electrical stimulation only when there is an important medical reason. That occurs now because anticonvulsant drugs are unsatisfactory in about 30 percent of patients; those who do not respond to the available drugs become candidates for surgery only if their epilepsy is of a particular type. For the others, there are no good answers.

A better anticonvulsant would improve the lives of millions of people. Yet opportunities to study language physiology would shrink at the same time, perhaps to the point at which research teams would disperse. It is true that other opportunities are opening up, such as improved x-ray techniques and positron-emission tomography (PET scan), but they do not allow manipulation of small regions of brain in the manner that has made the electrical stimulation technique so powerful: stimulation is a harmless probe, but it cannot be delicately applied through the thick skull. While it might be possible to figure out the workings of a clock by ingenious ways of looking through the walls of its case, the brain is far too complex a computer to analyze only by observation of its undisturbed spontaneous activities. Just as it is most useful to poke a finger in the clock's works to move certain gears in figuring it out, so it is with the neurophysiologist's use of electrical stimulation in figuring out language cortex.

Thus we may well have a restricted "window in time" during which our civilization can ethically study the more detailed workings of human language cortex, a window caused by knowledge in one area (neurophysiology) temporarily exceeding that in an adjacent area (neuropharmacology). By making the most of our opportunity to peer through that window and probe with harmless tools, however, we may greatly improve our understanding of part of that "human essence" of language and thought.

PROSPECTS

The *how* of human beings—every village gossip has been doing that since talking started. . . . It's the *why* of human beings you've got to understand. . . . Or else you'll be giving all your science to a mob of children. Whatever they do with it, they won't know why. We can never trust them. Unless they know the *why* about themselves, then everything in the world is like giving a child some poison and telling it to go and play in the kitchen.

C. P. Snow, The Search

There may be nothing new under the sun, but permutation of the old within complex systems can do wonders. As biologists, we deal with the kind of material complexity that confers an unbounded potential upon simple, continuous changes in underlying processes. This is the chief joy of our science.

Stephen Jay Gould, Ontogeny and Phylogeny

It is only when we can see the world as a ladder, and when we can see man's position on the ladder, that we can recognize a meaningful task for man's life on earth. Maybe it is man's task—or simply, if you like, man's happiness—to attain a higher degree of realisation of his potentialities. . . ." *E. F. Schumacher, Small is Beautiful*

Happiness goes like the wind, but what is interesting stays.

Georgia O'Keeffe

17

The Creation Myth, Updated: A Scenario for Humankind

> The fact that we want to know about origins may tell us much about the way our brains work. It emphasizes that we try to develop comprehensive schemes by which the antecedents of all events are explained and future events forecast. This characteristic separates us from animals almost as much as does language, to which of course our particular sort of model-making is related.
>
> *J. Z. Young, Programs of the Brain*

Anthropologists tell us that almost every culture, every primitive people, has a creation myth; for example, Adam and Eve. It is a story, repeated by word of mouth down through the generations, which attempts to account for human beginnings. Each creation myth is peculiar to the people who made it—and to their technology and folk wisdom at the time the myth was finally frozen by being written down.

The Cherokee Indians of the Great Smoky Mountains of North America talk of a creator who baked his human prototypes in an oven after molding them from dough. He fired three identical figures simultaneously. He took the first one out of the oven too early: it was sadly underdone, a pasty pale color.

Now, creators may not do things perfectly the first time, but in creation myths, their actions are always irrevocable—so the pale human was not simply put back in the oven to cook a little longer. It remained half-baked. But the creator's timing was perfect on taking the second figure out of the oven: richly browned, it pleased him greatly and he decided that it would become the ancestor of the Indians.

He was so absorbed in admiring it that he forgot about the third figure in the oven until he smelled it burning. Throwing open the door, he found it was black. Sad, but nothing to be done about it. And such were the origins of the white, brown, and black races of mankind.

It isn't just the Cherokees who gently proclaim their own superiority: creation myths always assign the starring roles locally. "Imagine the Lord talking French! Aside from a few odd words in Hebrew, I took it completely for granted that God had never spoken anything but the most dignified English," exclaimed Clarence Day in *Life with Father*. And it is hardly surprising that we create in our own image, as creation myths have more to do with evoking one's revered ancestors than with satisfying curiosity.

The scientific subculture of the last 200 years has also developed a creation myth, though it undergoes major changes every decade or two as new ideas and new facts compete. Some scientific fields, such as anthropology and human ethology, might even be said to be in the creation myth business—in the sense that their goals are to improve on creation myths, to assemble a factual account of human origins. To make an analogy between the intermediate, imperfect scientific accounts and the creation myths is not to disparage either but to recognize one of the ancient origins of our scientific endeavor. And the analogy is useful in that it leads us to examine our scientific tendency to assign the leading roles locally: to the clever, the technologically innovative, the communicative—indeed, to those whom we might like to consider as our revered scientific ancestors.

That our casting for our creation drama reveals some subcultural nepotism does not make the conclusion wrong. It may nonetheless be true that the success of primitive technology and language helped brains to become bigger and bigger. But it would be nice to have another competing theory too, just to see if a more mundane explanation will do as well.

And what, one might ask, would constitute a satisfying explanation of things uniquely human? After all, if we are going into the creation myth business, we should take the talents of the storyteller into account. Good stories should be like well-planned gourmet meals—not just a tasty morsel, nor necessarily everything possible, but a palatable full-course meal.

We might expect one course to be the ontogeny, such as the neotenic slowing of development so that the puberty alarm clock goes off and starts to arrest growth while many juvenile features are still present.

And another element would be a natural-selection story such as the throwing theory, which relates genetic changes and selection pressures.

The timing would need to be just so, relating what had happened in the 4.5 million years since the gorillas and chimpanzees went their own way.

We might expect a satisfactory helping of language physiology, illustrating some of the natural subdivisions for language in those areas that are uniquely human.

We would expect some linguistics, showing the natural subdivision of language learning and use (which should go nicely with the known language physiology—but don't count on it).

The *pièce de résistance* might be a nice mechanistic illustration of one aspect, something that you can really get your teeth into—such as a motor sequencing buffer being co-opted for language uses. Or maybe a special helping of human neuroanatomy, showing a unique set of pathways interconnecting hand motor strip and a temporal-lobe language area.

We would expect our satisfying spread to be garnished with other human *spécialités de maison,* such as right-handedness and gestures and left-armed infant carry.

We would expect it to provide a sound foundation on which to build social and aesthetic (music, anyone?) theories, a vantage point from which to better understand human variability and special skills.

Hours later, would it still be a satisfying meal, or would there be hunger pangs? Some restlessness, of course, is a good sign, as we would want it to be a productive theory—to lead to better meals next time, rather than just being a *fait accompli* lingering in our memory traces. In the spirit of the artists who intentionally leave a small imperfection in a finished tapestry, we must never presume to have truly perfected our explanation.

This would be a left-brain meal, if it could be assembled at all—though that time seems to be approaching. A whole additional smorgasbord would need to be constituted to cover the other lateralized abilities predominant in the right brain: body image, visual-spatial, prosody, emotional faces, and such. Any such story is merely a selection from many facts and theories—and it is always getting out of date. Still, trying one's hand at a creation myth is traditional, even if considered a sure sign of presenile dementia in some neurobiological circles.

We will pick up the story perhaps 15,000 million years after the Big Bang, during the most recent 0.2 percent of the lifetime of our universe. About 35 million years ago, according to the fossils and the molecular-clock evidence, the New World monkeys went their own way from the Old World monkeys—literally, because the South American tectonic plate drifted away from Africa, the valley between them widening about 2 centimeters every year during the breakup of Gondwanaland. About 20 million years ago, the hominoids less dramatically took leave of the Old World monkeys, merely ascending into the trees. These proto-apes developed a brachiating way of life, which gives us and the apes a distinctive torso and shoulder structure unlike the monkeys.

Then the gibbon split off at 12 million years, to become the most graceful of acrobats. The orang-utans split at about 8 million years. Gorilla, chimpanzee, and hominids all split off from a common ancestor about 4.5 million years ago, all this according to the molecular clock. Judging from the oldest hominid fossils, brain size back then was perhaps 400 cubic centimeters. There were an embarrassing variety of hominoid species by 1.8 million years ago, and in some the brain size was getting remarkably larger, starting the threefold enlargement that is such an extraordinary event on evolutionary time scales.

There are many levels of "cause." The proximate source of the energy that your car uses to accelerate is the gasoline in its fuel tank. The ultimate source of that energy is, however, the thermonuclear energy released on the sun. Among the many intermediate "causes," such as photosynthesis, is surely the sandstorm that buried the swamp so that its rotting vegetation cooked underground at just the right temperature for millions of years so that oil deposits were formed. You could thank a different cause every day as you turned the ignition switch but it would be a month of Sunday's before you started to run short of intermediate causes to thank.

To appreciate brain size arguments, one must be careful to understand that there are two kinds of "causes" in biology. Two extremes happen to correspond to medicine and to natural history, the two original endeavors which have gradually merged during the last century to become modern biology. The most familiar one to many of us is the functional cause: physiology is all about such mechanisms. When we ask about the "cause" of a bird's migration pattern, we can perhaps identify a change in hormone levels as the migration season approaches. Such a functional cause is also known as a "proximate" cause, in the sense of being the most immediate preceding event. Another kind of cause is distinguished in biology, often called the "ultimate" cause to place it at the opposite end of a spectrum of causes from the physiological ones. This is the cause of why one physiological mechanism exists rather than an opposite one, why a stay-at-home hormone rather than an emigration hormone, why one genetic program rather than another. Such ultimate causes have to do with ecological niches, natural selection, population dynamics—with the operation of a genetic program throughout an evolutionary time scale.

Evolution operates in part by selecting for useful variants in intermediate causes. And one is a developmental trend called neoteny. Neoteny seems to be a general feature of higher primates, and is seen in exaggerated form in humans. Adult primates look progressively more juvenile. Their faces, for example, tend toward the flatter nose and the larger eyes (relative to the rest of the face) of the young. Neoteny is achieved by a progressive slowing of developmental rates, so that it takes two years for a chimpanzee to grow as much as a monkey does in one year.

If we never grew up, of course, that would be the end of us. At some point, sexual maturity supervenes and growth usually slows to a stop soon afterward at some otherwise immature form by our ancestors' standards. It suggests at least two clocks keeping different time: one controlling growth rates, running slowly, and another controlling sexual maturity, slowed down but not quite as much. The pituitary gland's control of growth hormone is a traditional model for the first clock's mechanism (though one that has been much revised in recent years), and one candidate for the other clock is the pineal gland's control of melatonin.

But these are simply stand-ins for a more fundamental proximate cause, a matter that will eventually come down to alterations in regulatory genes. Indeed, it may be mostly minor changes in regulatory genes that separate us from the apes. We seem to share 99 percent of our DNA sequences with the chimpanzee and gorilla. Even our chromosomes, those protein backbones onto which the DNA strands are looped in a cell's nucleus like necklaces on a store's racks, look largely identical under a fluorescence microscope. Though the chimp has twenty-four chromosome pairs to our twenty-three, that seems to be because human chromosome No. 2 has enlarged to hold what is otherwise two separate chromosomes' worth of DNA in the chimp. The striking similarities suggest that a revolution didn't take place in the hominid line but rather something more subtle, a fine-tuning of sorts. Neoteny is just such a process. To adjust some developmental rate constants more than others would make the hominid phylogeny more analogous to manipulating interest rates in a complex international economy than to a political revolution.

It isn't just the shape of the face that results from neoteny. The adult head becomes larger, relative to the rest of the body. If you've ever seen a premature infant, you may have been shocked to discover that it was nearly half head (indeed, neurologists and neurosurgeons are always being consulted by premature parents worried about the possibility of hydrocephalis). In normal newborns, perhaps a fourth of the body is head. And while the head grows quite a lot in the first postnatal year, growth of the rest of the body outstrips it. If relatively bigger adult brains are a good thing, as determined by selection pressures, then neoteny is a genetic trend that could be reinforced. About one-seventh of the adult body is head—but it used to be even less.

There are some problems with retention of juvenile features, not the least of which is getting the infant head through the relevant portions of the adult female pelvis. One solution seems to be a somewhat premature delivery date, with the infant born even more helpless than usual, even before the skull is fully formed (that soft spot called the fontanelle). Some would consider much of the infant's first postnatal year to be equivalent to the last months of gestation in other primates. Another problem is the loss of body hair: juvenilized adults may have less body hair, an important matter to primate infants who cling to maternal hair for transportation while mother goes about making a living. And with the infant's state of development at birth becoming even more helpless, mothers must become even more skillful at carrying while gathering.

Problems, yes, but also advantages: more brain, relative to the body it must run. And a more flexible brain at that, retaining such juvenile features as curiosity, play, and a willingness to try out new things. Which may have been far more important than bigger brains per se. Juvenile primates are preprogrammed to learn by imitation of others (a feature which is most impressive in human children who, unless deaf, will automatically learn any language to which they are exposed). The long childhood presumably evolved because it made for more adaptable adults, more capable of behaviorally molding themselves to the environment in which they found themselves rather than just into the traditional niche occupied by their ancestors. This is handy for adapting to a changing world (the Pliocene and Pleistocene) and for expanding the range, to occupy a variety of habitats outside the tropics.

It is thought that upright posture evolved from this background much earlier, before 3 million years ago, perhaps from a hairless mother's need to carry her more-than-usually-helpless infant. But if infants can be carried, so can food. Instead of consuming it on the spot, food can now be carried back to a nest. In the same manner that birds and wolves bring home the bacon, so too could an omnivore hominoid. And not just a mouthful at a time (or a stomachful, for those species that regurgitate food for their young), but an armload, particularly important for foods lower in calories than meat.

And pairing, also common in birds and wolves, could have become an important survival trait: there would be two adults feeding the young, rather than just the mother. Many species manage without pairing, but for higher primates, a little help is most important. Even the modern-day great apes, who stick to lush tropical forests where making a living is easy, are just marginal when it comes to reproducing their kind (even without the depredations of human civilization). The long childhood means that chimpanzee births are spaced perhaps five to six years apart (a young chimp may nurse for four years), as the mother simply cannot effectively care for more than one child by herself. Even with a fifteen- to twenty-year reproductive period, getting two offspring to survive until reproductive age is hard for a mother to achieve, given the infant mortality. This extreme parental investment in a few offspring is the far end of "K-selection" reproductive strategies; rather than using a shotgun approach like frogs, such species specialize in the well-aimed single shot. It is thought that the great apes may have overspecialized themselves toward extinction.

Only with the additional investment of the father's energies does the shorter human spacing between children become possible. But how did that come about? Like cooperation, paternal investment may be good for the species but not to the immediate advantage of any one individual (a male reproductive strategy for maximizing offspring known as "love them and leave them" achieves quantity at the expense of quality). Chimpanzee males, for example, probably do not know which infants they sired; while such behaviors as guarding the troop are useful to all infants, the chimp male probably cannot selectively boost the chances of some particular infants so as to help perpetuate his particular genes.

So how was some male help recruited? It is thought that the continuous female sexual receptivity, quite different from primate estrus, is a hominid characteristic related to solving this K-selection problem—essentially, attracting the male to stick around and, just incidentally, help care for his offspring. Such reproduction considerations, as Owen Lovejoy as pointed out, may well have been what set the stage for the upright posture and carrying. And they may be one reason that the hominid line has flourished better than the chimpanzee and gorilla line.

Thus the background to the great encephalization seems likely to include neoteny, upright posture, and pair-bonding. Encephalization must be judged by comparison to other animals, whose brain size varies with body size. A log-log plot of brain size versus body weight has a nice straight line running upward at a slope of two-thirds; double the adult body weight, and the brain will enlarge about 59 percent just to run the larger body. Most vertebrate species plot somewhere close to this line, and the neotenized monkeys and the great apes are somewhat above it. But humans are way off the line with triple the brain volume of comparably bodied great apes.

There is no mention of handedness, language, or hunting in this sketch of our ancestors millions of years ago. Yet the picture differs from our present-day cousins, the great apes, in several aspects: upright posture, pair-bonding, and the lack of both hair and estrus. Compared to chimps or gorillas, ours was a better base from which to overcome the extreme K-selection of the higher primates (though it seems unlikely that anyone rationally considered this little problem in ecological economics!). Perhaps our ancestor was capable of living in more widespread habitats than present-day great apes, and perhaps there was some difference in the extent of neoteny. But there is nothing to say that this ancestor was smarter or better-spoken than chimps and gorillas.

The oldest of the fossil skulls from Hadar, in the Afar triangle of Ethiopia where "Lucy" was found, is more than 3 million years old and suggests a brain size between 370 and 450 cubic centimeters, right in the chimp-gorilla brain size ballpark. But something happened 2 million years ago: the hominid brain began a sustained growth and tripled to attain its present 1400 cc average size. Leading, of course, to the inevitable threshold arguments about "how big does a brain have to be if it is to have human capabilities?" which I shall spare the reader. Crossing this Rubicon (alleged to be between 700–800 cc) borrows another doubtful metaphor from physics, that of the chain reaction which, once a certain critical mass is achieved, takes off on its own. Personally, I'd say that 473 cc was the important threshold for brain size (I will break my rule against nonmetric units only long enough to note that this achieves a pint-size brain).

We also differ from the great apes in the extent to which we make use of hunting, language, and tools. General intelligence is often invoked, but that is harder to evaluate and it tends to turn the question around. We should ask first what encouraged that bigger brain up to the point that size became a virtue unto itself (if it is—certainly in present-day human adults, there is no correlation with "intelligence" as brain size varies normally between 1000 and 2000 cc).

And so we ask what, given the background sketched out above, is there about hunting, language, or tools that especially interacted with brain size? Other animals hunt and carry meals home to a family. Other animals use tools and even make them to a pattern. And chimps are certainly clever enough to make use of many elementary tools. Other animals use elaborate systems of vocal communication. What, in any of these, is there that could become so successful? Successful in making humans the most widespread and adaptable of animal species (hopefully we will retain this title, rather than passing it on to some radiation-resistant insect that thrives on a radiation-resistant grass), successful in selecting for bigger brains?

The usual picture pieced together (the "hunting hypothesis") has all three interacting during group hunting of large game animals: the band of hunters using communication and strategy to corner a prey animal, then using tool-sharpened spears to pierce its tough hide, cutting tools to carve up the animal, and scraping tools to save its hide for clothing. Which we might call the Sergeant-Pepper's-big-game-hunting-band metaphor.

But is it reasonable to look upon this social-hunter metaphor as representative, back when brains were a third their present size? Solitary hunting as an adjunct to gathering seems a more reasonable first step, using strategy to succeed rather than size or speed. Tool use, as when chimps hammer open nuts with a handy rock, seems an earlier step than spears. And the spear throwing used by the hunting band must itself have precedents, intermediate between that advanced form and the unaimed throwing used by chimps in threat displays.

The communication needed to corner a prey, on the other hand, may be no greater than that used by chimps when they cooperatively corner a small monkey for a meal: they become quiet, take up alert postures, and move to block escape routes. They obviously accomplish a lot with body language (much as do modern-day hunting bands, even those encumbered with technology such as army patrols quietly probing enemy defenses). So elaborated verbal communication for hunting seems unlikely to engage selection pressure for bigger brains, at least when starting off from the biological base that existed a few million years ago. Body language seems sufficient for the task.

Body language, like its elaboration, the sign language of the deaf, does not have a prominent syntax, an ordering of signs that is important for deciphering the meaning. For the hominid precedents to our sequential ordering of otherwise meaningless sounds used in verbal language, we perhaps need a very sequential behavior of some kind to provide the foundation, the right neural machinery.

In more than one chapter of this book, the throwing theory has been put forward to illustrate how small-game hunting could have suitably engaged the selection pressure ratchet in favor of both sequencing machinery and bigger brains. While intended as more than a metaphor, the throwing theory is but one example of a "fast track" possibility, standing in contrast to the generalized intelligence theory which says that brains beget bigger brains because intelligence is good. To say that throwing provides a fast track is not to deny other important elements of hominid evolution such as the carrying basket or the two-parent family; it is only to focus on a feature that could be associated with rapid encephalization. Whether or not throwing turns out to be the fastest track of all, something mundane may be primarily responsible for our most prized human abilities.

This should come as no surprise, but it does. Our present-day world is full of examples of sophisticated abilities arising in a short time from humble beginnings. In less than forty years since computers started pointing antiaircraft guns and breaking ciphers, the same types of circuitry have given us word processors and sophisticated examples of artificial intelligence. Not to mention all the smart toys for the kids. The newspaper the other morning had a humorous quote: "One humiliating thing about science is that it is gradually filling our homes with appliances smarter than we are."

New uses for old things—that is a theme in biological evolution as well as in modern technology. Feathers made good thermal insulation; when reptilian forelimbs had enough such thermal insulation, it probably became possible to glide downhill. Which, because it made catching flying insects easier (or some other utilitarian reason), promoted even more varieties of feathered animals. To identify the graceful flight of birds with more mundane beginnings is not to denigrate it but to better appreciate how evolution builds sophisticated capabilities step-by-step.

Humans are the product of a hominoid ancestor, shared with the great apes, and a unique series of selection pressures and ecological opportunities. For about the first 3 million years of separate evolution, both hominids and apes seem to have remained confined to Africa. But for much of the Pleistocene, hominids expanded outside the tropics. They were especially exposed to the selection pressures accompanying the changes in climate. Each Ice Age has also served to geographically isolate tribes of hominids, probably promoting inbreeding and speeding specialization on the fringes of the main population. The arctic ambience every 100,000 years has left behind an amalgam of abilities, producing a capital investment on which we have been living ever since. Increased skills in hunting, both in herding strategy and in projectile predation, surely occurred. And the improving rapid motor sequencer used for throwing could have also been used for speech—perhaps at special off-hours low rates.

Democritus and Aristotle thought that chance and purpose were the only two alternatives to explain life on earth. Scientists used to focusing only on proximate causes often continue to get hung up on that traditional dichotomy, even some Nobel-winning molecular biologists (and, alas, neurophysiologists as well) who pontificate on origins. But for more than a century, there has been another alternative: the Darwinian Two-Step (variations, then natural selection, alternating back and forth), an ancient dance where the changing environment automatically shapes new species from old, producing a stratified stability, a hierarchy of proximate causes.

And more recently, we have learned about the occasional sidesteps—how novel uses are made of old things, so that unexpected things (like flight) arise from prolonged environmental selection for something else (like keeping warm). The human capabilities that allow us to read and understand this sentence—neural machinery that must have existed for a long time before writing found a new use for it a mere 5000 years ago—had an evolutionary history too, one that involved many stepping-stones and unexpected sidesteps. Surely one of the highest uses of our intellectual abilities is to understand how they evolved.

APPENDIXES

APPENDIX A

Did Throwing Stones Shape Hominid Brain Evolution?

WILLIAM H. CALVIN
University of Washington

Reprinted from *Ethology and Sociobiology* 3:115–124 (1982).
Ethology and Sociobiology 3: 115–124 (1982)

52 Vanderbilt Ave., New York, New York 10017

Early hominid evolution may have involved an interaction between lateralization to left brain of rapid motor sequencing (e.g., right handedness) and its selection via one-handed throwing of stones at small prey. Since a more redundant sequencer should permit faster orchestration of muscles, faster (and hence longer range) throws could have selected for encephalization. Secondary uses of the enlarged sequencer may have included tool-sharpening and manual gestures. Because an oral–facial sequencing area just below motor strip forms the core of modern language cortex, there may have been a common origin of handedness and language in redundant sequencing circuits selected by throwing success.

Key Words: Aphasia; Brain; Cortex; Encephalization; Ethology; Evolution; Fire; Gene; Generalization; Gesture; Habitat; Handedness; Hominid; Hunting; Language; Lateralization; Linguistics; Maternal; Motor control; Neoteny; Opposable thumb; Physiology;

Pitching; Pleistocene; Population blush; Predation; Redundancy; Selection; Sequencer; Threat; Throwing; Timing; Tool.

Introduction

Were there any early, especially significant events in the evolution of the hominid brain? We normally think of a gradual drift of small improvements, with language proving incrementally useful to social predators when hunting woolly mammoths, or visual–spatial specializations being useful to the cooperating animal for recognizing cheaters' faces. But were there jumps as well as drift, and might a major aspect of hominid brain evolution have been due to a single genetic (or cultural) invention? If so, what were the strong selection pressures which shaped the genome?

The hypothesis presented here is that rapid motor sequencing for hand and arm (Kimura, 1976) was one of the earliest of hominid lateralizations—and that it strongly interacted with the cultural invention of one-handed throwing of stones at prey, a behavior that is surely selectively advantageous (Darlington, 1975). This would have allowed many new prey to be eaten, such as rabbits, squirrels, and birds, which were too fast to chase but which remained stationary provided that a hunter kept his distance. Perhaps quite suddenly on our usual time scale, hominid population expansion could have then occurred into many new habitats having minimal forage, just from this single invention. Not only does one-handed throwing have this "new niche" aspect but its faster-is-better aspect has an unusually strong potential for enlarging a center in one hemisphere (Calvin, 1982).

If such neural machinery for sequencing was also used for the oral–facial musculature (near the hand in the motor strip), it could have been the scaffolding upon which language was built. Recent mapping evidence shows that the core of the human language cortex is an area specialized for oral–facial sequencing (Ojemann and Mateer, 1979). A common origin could explain why one of the modern manifestations of the motor sequencer, the tool-use and throwing preference known as right handedness, is usually (though not inevitably) associated with the same left cerebral hemisphere as language.

Genetic Tendencies and Natural Selection

In the beginning—which for the present discussion would likely be back in the Pliocene or late Miocene, perhaps two to eight million years ago—it is convenient to postulate two genetic tendencies in the background: a novel combination of genes which led to relatively larger brains [Gould (1977) makes a strong case for neoteny via retarded maturation retaining the juvenile brain/body size ratio], and another that led to a lateralization of the neurons involved with orchestrating rapid muscle activation sequences other than locomotion [some 15%–25% functional left–right asymmetries are seen in motor aspects of even rat brains (Glick and Ross, 1981)]. Such simple genetic effects would hardly be macromutations; the throwing genes may be nothing more than these two trends.

The present paper explores one "fast track" hypothesis for some post-Pliocene hominid traits [for some essential background, see Lovejoy (1981) for a recent synthesis of the Miocene and Pliocene fossil record with demographic and social theory]. One has, after all, to account for the prodigious tripling of hominid brain volume since the Pliocene (Holloway, 1976) and the extraordinary extent of hemispheric lateralizations, not just speciation per se and the abilities unique to modern man. Precision throwing, made possible by the cultural invention of projectile predation, generates an unusually strong and sustained selection pressure, has a one-sided aspect that can promote lateralization and handedness, has a major use for redundancy and thus bigger brains, and promotes sequencer neural circuitry of the type needed for both receptive and expressive aspects of language. The throwing theory is thus primarily based upon natural selection and secondary uses (preadaptations) or, to use a retrospective rather than prospective term, exaptations: (Gould and Vrba, 1982), not upon genetic revolutions. Though the effects discussed are not inherently stepwise, they should operate as well under punctuated evolution as under Darwinian gradualism: selection may well operate most effectively and rapidly in the small, isolated, inbreeding populations where speciation more readily occurs (Mayr, 1963; Stanley, 1979).

From Casual Predator to Action-at-a-Distance Predator

It is not obvious how the chase, or other primate food strategies such as foraging, could have evolved into throwing. I assume for the moment a line of invention involving chimpanzeelike opportunistic predation, though other invention sequences may be equally likely.

Monkeys have been observed to drop coconuts from a tree to crack them open, in a manner reminiscent of seagulls dropping snail shells. Chimpanzees are said to throw rocks at the skull of a dead monkey, also in aid of extracting a delicacy food from its interior (the original *pièce de résistance?*). One can imagine intermediate steps linking such a battering throw to the one-handed throwing useful for hunting.

One hazard of being a predator is injury by a prey animal that can still thrash around; the fitness of lions is not improved when they encounter a wildebeest horn. There were probably hominids that did not wait until after the main meal to attack the cranial vault with a rock—and hence discovered a useful way of permanently disabling their prey while standing outside the range of their wounded prey's teeth: throw first, eat later. Action at a distance is far preferable to mixing it up, as when chimpanzees break a monkey's neck manually or swing the monkey by its legs into a tree trunk (Teleki, 1973). From braining a downed prey to throwing the stone to down the prey in the first place is a simple invention sequence exposed to natural selection. Another pausible line of invention could have started from threat displays (see van Lawick-Goodall, 1971, p. 216) with threatening throws being converted into precision throws, an idea going back to Hall (1963) and elaborated by Hamilton (1973); at least in the savanna where game density is high, one can imagine a hominid initially using such throws to drive a predator away from his kill. A chimpanzee has been seen to throw a melon-sized rock to scare away adult bush pigs who were protecting a piglet from predation (F. X. Plooij, cited in McGrew, 1979). Though of obvious use for intraspecies aggression and defense (and chimpanzees have been observed to throw large rocks during brutal attacks on other chimpanzees: Goodall et al., 1979), throwing's everyday uses for plundering and hunting game would seem more exposed to selection.

The chimp's two-handed throw (e.g., van Lawick-Goodall, 1971, appendix C) is of limited value for predation unless the throwing distance is extended. One-handed overhand throwing (van Lawick-Goodall 1968, p. 203), with its windup allowing axial and elbow angular momentum to also contribute to the stone's velocity, does that. Its invention need not have coincided with the lateralization of rapid motor sequencing. But once they both existed, there would be a tendency to use the hand and arm opposite to the better sequencer.

Selection Shaping of a Sequencing Center

A one-handed skill would interact with a unilateral neural specialization; though not a unique event in evolution [e.g., the asymmetrical shrimp Alpheus with its one large claw having a specialized snapper-type motor program (Wilson and Mellon, 1982)], the throwing aspect would seem to favor rapid evolution. The back-and-forth interaction between improved throwing skill and better neural machinery could progress at an usually rapid rate because of three factors: what might be called the "fast ball" effect, the "new niche" effect, and the "two-for-the-price-of-one ratchet."

Outside of saccades and locomotion, most movements do not have an internal imperative for speed, e.g., grooming or picking fruit. But throwing does: faster and faster is better and better. The primary advantage of speed is throwing range, as it is a function of the velocity at which the rock leaves the hand. A second advantage of speed is that a fast projectile gives the prey less time to react. Furthermore, the "stopping power" of a projectile is a matter of transferring kinetic energy: while this is proportional to the projectile's mass, it is also proportional to the square of the velocity. Approximate ninefold improvement occurs when one triples the speed. The lighter-but-faster strategy (even some army rifles now use a small caliber bullet backed by lots of gun-powder) would have meant that the animal using a rapid one-handed throw of a small rock would have had the advantage over a more symmetric-brained animal slowly throwing a large rock with two hands. Hence faster-is-better—but what are the brain correlates of faster?

Bigger Brains and Precision Timing

While it seems likely that many slow actions such as grooming are coordinated by an ad hoc committee of multipurpose neurons residing in many regions of the brain, a concentrated neural "center" would have substantial advantages where speed is of the essence (more direct connections to motor neurons; minimized impulse conduction time; neurons in centers tend to be more committed to particular roles, reducing the time needed to resolve ambiguity). During about 1 sec of throwing time, the desired trajectory of the rock must be translated into a muscle sequence, orchestrating the times at which the various muscles are brought into play to accelerate and then precisely releasing the rock within a very narrow "launch window."

The escape reflexes of fish and crustacea illustrate one way to orchestrate a rapid multi-muscle sequence: they typically utilize a single giant neuron whose axon gives off branches to many motor neuron pools. The axon propagation delays allow a single impulse to trigger different motor neurons at unique times, thus flipping the tail in a coordinated (but stereotyped) manner. The hominid sequencer, however, would need not only rapid orchestration but delays that could be precisely graded to correspond to various trajectories. To play out such a "motor tape" with submillisecond precision seems likely to require many coordinating neurons to achieve adjustable-but-precise delays between different muscles.

One way to improve timing precision is through redundant timers, for example, on the 1831 voyage of the *Beagle,* 22 chronometers were used to precisely determine longitude from local noon. Models of circadian oscillators show that an effective way to achieve precise timing with imprecise individual neurons is to use an ensemble of many redundant neurons (Enright, 1980): precision increases with the square root of the number of neurons if they are arranged in a simple circuit. Calvin (1983) shows that overhand throwing has a shrinking launch window, such that the time of rock release must become 8-fold more precise as the throwing range is doubled from a beginner's throw; were redundancy used to accomplish the precise launch, 64-fold more timing neurons would be needed to double the throwing range from 4 to 8 m. The precision required is not a direct kinesthetic measurement but a constraint imposed by the physics: doubling the range not only halves the angle subtended by the target but also typically requires doubling the "muzzle" velocity and thus shrinking the time scale. The conclusion that a 4-m throw is already operating near the limit of the physiological cellular noise is based upon long experience with spinal motoneurons (for example, Calvin and Stevens, 1968); that 64-fold redundancy is used to bridge the noise gap is simply the only known possibility at this stage, but one for which the requisite circuit and numbers have been worked out (Enright, 1980). The conclusion in Calvin (1983) is that some bridging mechanism is required to meet the constraints imposed by Newtonian physics and that redundancy could do the job.

While a motor-sequencing center might enlarge at the expense of other adjacent functions, the easiest way to get more sequencer machinery in a concentrated center would be to simply enlarge the next-generation brain so that there was some extra uncommitted cortex surrounding the center which the sequencer could appropriate. The enhanced throwing abilities could provide a strong selection pressure for any encephalization trends which multiplied the numbers of redundant timing circuit neurons in sequencer cortex.

This use of redundancy is quite different from the usual concept of a "fail-safe" backup, a redundant system which is exposed to selection pressures only when all others fail. Each successive backup system is less exposed to selection and one would expect the growth curve to flatten. This is in considerable contrast to redundancy for precision timing in throwing, where exposure continues and each additional meter of throwing distance makes square-law demands on redundancy for the requisite timing, suggesting a sustained growth curve.

The "New Niche" Effect

The postulated genetic tendencies—bigger brain and lateralized sequencer—could thus interact strongly with the cultural invention of throwing stones at prey. One-handed throwing, having a greater range or accuracy due to a somewhat larger brain, would conserve those genes through increased fitness. An ability to eat small mammals and birds would allow a foraging population to expand into quite different habitats. The throwing animal would essentially have discovered an empty ecological niche for action-at-a-distance predators [just as there is currently an empty niche for a bacterium which can digest nylon (Boulding, 1978)]. Population explosions are the traditional reward for first finding such a new niche.

Once invented, precision throwing would be on a faster-is-better curve for some time, the animals with faster throwing machinery having an ever better ability to feed themselves and their offspring. Since throwing range is proportional to velocity, and since "stopping power" goes up with the square of the velocity, the increased range would not only allow hunting the more wary animals but also the bigger, harder to stun, animals. The baseball pitcher's 140-km/hr fast and the outfielder's 100-m throwing range, indicate the current plateau of throwing development. The bigger-is-faster-is-better aspect of a sequencing center suggests that throwing skills could have continued to select for bigger brains from the Pliocene to recent times, aided by such supplementary inventions as the javelin, throwing stick, and sling.

The Lateralization Bonus

Since the (here postulated) genetic influences would produce global enlargement of both hemispheres rather than specific local hypertrophy, selection pressures favoring enlargement of a left hemisphere sequencer would give rise to more-or-less-symmetrical enlargement of the right brain as well. This two-for-the-price-of-one effect would likely allow the incidental improvement of other functions—such as the visual–spatial functions that seem to have settled in, opposite language. Visual–spatial skills are also generally useful for hunting (without accuracy, expanded throwing range would count for little).

The left/right opposition of sequencing and visual–spatial functions is probably a special case, in that the ratchet can be driven from both sides by the same strong selection pressure, precision throwing. But one can imagine examples where one skill could "bootstrap" another unrelated skill, thanks to lateralization (was musical ability a bonus of a contralateral specialization, or selected on its own merits? Or neither?). This can be seen as an advantage of lateralization, provided that extra neurons are useful contralaterally. Indeed, the enlargement of the rest of the cortex—the encephalization of hominids relative to other mammals during the past few million years—could be largely secondary to the bigger-is-faster-is-better selection pressure on the unilateral locality involved in rapid sequencing.

Note that "throwing genes" are not genes-specifically-for-throwing but the genes-conserved-by-throwing-success, such a sequencing lateralization genes and bigger-brain genes—perhaps merely slower-maturation genes [which retain juvenile plasticity in adults as well as the larger brain/body size ratio and the forward location of the foramen magnum of juveniles (Gould, 1977)]. To modulate some developmental rate constants more than others seems more like fine-tuning than macromutation; "throwing genes" could turn out to be more analogous to manipulating interest rates in a complex international economy than to a political revolution.

The Tool-Sharpening Bonus

Besides extra space being created for functions not under strong selection themselves, there might have been carryover: the rapid motor sequencer itself being useful for skills (such as music) other than those that first shaped it through natural selection.

Using one rock to chip another rock—to shape a chopper or scraper, or obtain a sharp flake—also uses a rapid unilateral movement sequence: one typically holds the tool-to-be with the left hand and then uses the right hand to strike a sharp, well-aimed blow with another rock. Like overhand throwing, this uses a rapid extension of the elbow joint for the major motion. Speed is again important because of the square-like kinetic energy consideration: a fast blow carries much more energy with which to break open the cleavage planes in the target rock (brute force may not be essential to a diamond expert, but here we are concerned with origins and amateurs). Faster brains can flake harder rocks which hold a sharper edge. Because it too involves rocks and rapid unilateral elbow extension but is less exposed to selection, tool-sharpening seems a natural sequel to throwing, just as fire-starting would be a likely tertiary development (via sparks which fly when striking hard rocks together). Darlington (1975) notes that making the using pebble tools is a more complex behavior than throwing unworked stones. The nut-cracking practiced by female chimpanzees (Boesch and Boesch, 1981) with natural hammers illustrates a possible antecedent of tool-sharpening (chimpanzees flaked their rocks while hammering the hard Panda nuts in four instances); indeed, one could also argue that such hammering selected for the neural machinery also useful for throwing, with throwing then serving to further select for left brain sequencing and right-handedness (Calvin, 1983). Annett (1970) showed that human right-handedness is strongest for the ballistic movements (hammering, clubbing, throwing), not fine movements (such as threading a needle).

Was the opposable thumb—essential for making many tools—also first shaped by throwing? The spatially precise pinching grip, so useful to gelada baboons for seed foraging in the plains, could have evolved into a temporally precise opposable grip by the necessity of holding the stone during the throw and then letting it slip at exactly the right millisecond. Although tool-use and tool-making are no longer the unique provence of hominids, precision throwing may have facilitated a wider, more versatile exploitation of tools by providing better sequencing machinery for hand and arm.

Concentrating the Throwing Genes Through Maternal Throwing Skills

Before further exploring the secondary uses of a lateralized sequencer, consider a hominid subpopulation after a population blush, now isolated in a new habitat with modest forage but many birds and small mammals: climatic alterations serving to reduce forage would then put a premium on hunting skills. In addition to a general tendency to select for the more efficient hunters in this overextended population, throwing skills could also influence selection via some successive-generation interactions where acquired parental skills serve to conserve the throwing genes in their offspring.

Predator–prey chases present the unaided mother with the difficult choice of carrying the infant along for the ride (which is tiresome) or of leaving the infant temporarily unattended (which increases infant mortality via accidents or predation). A chase might also increase spontaneous abortions. The mother endowed with throwing genes could better throw rocks at nearby small mammals and birds. Her success in this regard would have a quite direct effect on conserving the throwing genes, as it would enhance both her offspring's food and protection. Her throwing skills would affect the chances of passing on throwing genes to grandchildren. Such "Lamarckian" effects are not unexpected from a cultural practice such as throwing; what is more remarkable is the suggestion that an animal cultural practice could drive encephalization so directly. Since the key invention is cultural, with selection acting upon genetic tendencies, the throwing theory provides a major example of genes "tracking" culture (Cavalli-Sforza and Feldman, 1981; Lumsden and Wilson, 1981). So embedded are the genotypes by now that infants may practice ballistic arm movements (hammering, throwing food) well before they begin to talk.

While male behaviors such as guarding the troop may be generally useful, the skills of a philandering male do not usually boost the success of his own offspring more than others—absent kinship, of course. Although one scarcely needs another reminder in this century which started with the European-American inventors Tesla and Marconi and has progressed to Japanese transistor radios, this serves to emphasize that origins should not be judged by present-day hypertrophy: the first star pitcher may have been a mother.

Indeed, an early female hominid specialization for throwing would explain a puzzling finding: modern females tend to carry infants on their left side (Salk, 1973; Lockard et al., 1979), thus freeing their right arm. This is not merely a general adaptation to right-handedness as males (as a group also overwhelmingly right-handed) exhibit no hand preference for infant carry; mothers separated from infants after birth for some days do not subsequently adopt the strong left-armed preference of mothers who begin carrying their infants immediately after birth (the infant crying unless it can hear the mother's heartbeat, to which it became accustomed *in utero,* is the major theory for the left-armed carrying). The strength of the preference (3 : 1) cross culturally suggests that it may have deep roots in hominid phylogeny.

One hypothesis to be considered is that the maternal preference for left-sided infant carry is what promoted female hominid right-handed throwing, the success of which could have played a unique role in promoting encephalization and handedness.

The Sequencer as Scaffolding

Generating Gesture, Grimace, and Grammar

Another bonus which might result from the sequencer would be hand and arm gestures—which brings up the issue of language. Although it is natural to suppose that verbal language evolved out of the usual species-specific vocalizations, there is a problem with that: the cortical specializations for such vocalizations in monkeys are near the corpus callosum (Sutton, 1979), far from the peri-Sylvian language area of man. The gestural theory of language origins [which was actively debated in the 18th century (Hewes, 1976)] has been advanced as one alternative. Instead of snarls or signs, might a third alternative—sequencing—provide a more fundamental foundation for language cortex (Kimura, 1976)?

Recent mapping experiments in human language cortex by Ojemann and Mateer (1979) have shown that an extensive central core of the left language area is devoted to the sequencing of oral–facial musculature and to the discrimination of phonemes; surrounding this core is a patchwork of areas involved with semantics, syntax, and short-term verbal memory.

Though oral–facial rather than hand–arm sequencing, those experiments grew out of the observations by Liepmann (1908) and Kimura (1976, 1982) on manual sequencing tasks in patients with left hemisphere strokes. Such patients have difficulty mimicking a sequence such as turning a key, then rotating the doorknob and pulling/pushing the door—with either hand. Mateer and Kimura (1977) then showed that left (but not right) hemisphere stroke patients had similar difficulties mimicking a series of oral–facial sequences—with either side of the face. It is the oral–facial sequencer that has been further localized by stimulation mapping in awake patients during epilepsy operations (Ojemann and Mateer, 1979; Ojemann, 1980; see also Calvin and Ojemann, 1980, Chapter 4), but the manual sequencer is presumably (stroke evidence is less precise) nearby, just as the face is near the hand on the motor strip above the Sylvian fissure [though the issue is complicated by its interrelation with parietal lobe apraxia theories (Kimura, 1979, 1982)].

Certainly the tendency of both right-handedness and language to be represented in the left hemisphere is compatible with a common origin in a lateralized motor sequencer. Though it is difficult to evaluate innate throwing abilities in a culture where practice effects favor males, there may be a tendency for modern males to be more natural pitchers—and for them to have language more strongly lateralized than females [males are far more likely to suffer aphasia than females having comparable left hemisphere damage (McGlone, 1980)]; perhaps a sequencer common to language and throwing is more strongly lateralized in modern males.

The Sequencer as a Point of Departure for Other Hominid Cortical Specializations

That toolmaking, hunting and language have something to do with one another and with handedness has long been suspected (cf., Montagu 1976); the throwing theory merely identifies a single neural mechanism that could tie them together and proposes a related strong selection pressure which could shape it.

The sequencer seems likely to be a core resource for music, throwing, tool-sharpening, manual gesturing, and such rapid laryngeal–oral–facial movements as in language expression. Some auditory perception and memory abilities (Efron, 1963; Tallal and Newcombe, 1978; Ojemann and Mateer, 1979; Sherwin and Efron, 1980; Tallal and Schwartz, 1980; Bradshaw and Nettleton, 1981; Ojemann, 1982) are also strikingly dependent upon timing and sequencing (consider the transitions which characterize the difference between the sounds "L" and "R"). While, at first glance, the receptive aspects of language are seemingly unrelated to "motor" skills, there are in fact a series of intimate interrelationships (Ojemann, 1982) dating from the motor theory of speech perception (Liberman et al., 1967).

Elaborate communication can take place with gesture, facial expression, and eye contact—and it need not rely upon a sequential ordering of elements (see, for example, Patterson and Linden, 1981, p. 112). But much of the power of spoken language systems derive from their use of word order—syntax allows a few dozen individually meaningless phonemes to be sequenced to form a very large number of meaningful messages. Thus sequencing may not be fundamental to language systems in general but is rather an attribute to one of the most productive systems. Cortical sequencing specializations have been indistinguishable from those for precision timing, though there is no a priori reason why these two temporal characteristics need be related. It could, for example, relate to a need for a "holding buffer" in which the individual commands in a motor sequence could be preloaded and then rapidly and precisely emptied ("get set" and "go")—and that might secondarily be useful to temporarily hold the received components of a long complicated sentence (such as the present one) while the rules of language analysis are recursively applied to its phonemes, its phrases, and then to the whole sentence itself.

We presently know little about issues of common-use versus separate sequencers (except via such experiments as trying to pat your head at the same time as rubbing your stomach): Lomas and Kimura (1976) show impairments of sequential finger and arm movements if speaking concurrently. It is, moreover, noteworthy that the motor strip layout goes from hand to face to larynx—suggesting a possible order for developing secondary access to a sequencer initially selected for hand-arm connections, for example, manual gesturing, followed by sequenced facial expressions, followed by elaborated vocalizations (but gesture–grimace–grammar could be a parallel development as well as a serial one, with similar sequencing circuitry being adopted in each cortical area). The many-hundred-fold redundancy useful for precision throwing occupies those extra neurons only during an actual throw; their timing/sequencing abilities might be utilized for other purposes in between throwing occasions in the same way that reptilian feathers for thermal insulation are thought to have developed secondary uses for bird flight (Ostrom, 1974; see also Gould and Vrba, 1982).

The throwing theory speaks to improving hominid visual–spatial (especially depth perception) skills, though with less detail than it speaks to selecting for rapid motor sequencing and its lateralization. Certainly, the throwing theory does not specifically account for some other features of modern visual–spatial localization such as the right posterior temporal lobe's specialization for judging the emotional expression on other people's faces (Fried et al., 1982); socialization and cooperation theories suggest effective natural selection pressures for such an area. The throwing theory primarily focuses upon the circumstances initiating the great encephalization; while predicting that a strong selection for bigger brains (more redundant motor sequencing and throwing's visual–spatial counterparts) could continue through until agriculture supervened to decrease throwing's exposure to selection pressures, it does not specify how other functions will be augmented other than noting those that can make secondary use of the precision sequencer machinery such as tool-sharpening and language. Improved functions from increased neuronal numbers is an assumption, but probably a reasonable one.

Generalization Versus the Fast Path

The enlargement of the hominid brain has long been thought to correlate with other hominid traits such as tool-use and bipedalism; most of these traits are now known to be neither exclusively hominoid (not even projectile predation: archer fish spit with great accuracy at flying insects) nor to parallel the hominid encephalization [having been present well before the great encephalization started several million years ago (cf., Lovejoy, 1981)]. But other traits such as our sequential language elaborations may have indeed paralleled encephalization.

Most workers have related language and intelligence to general, rather than specific, developments. Much emphasis has been placed upon intense socialization and upon the role of generalization: Lovejoy's (1981) theory relating bipedalism for carrying, pair-bonding, and the continuous human female sexual receptivity to a demographic strategy provides some specifics for the intense socialization concept up through the Pliocene.

Generalization has a similar appeal for explaining the Pleistocene developments: the tautology hazards notwithstanding, we recognize the mutual interrelationships between encephalization, intelligence, and language—and usually proceed to hypothesize that they, in effect, bootstrap one another ("brains beget bigger brains" might serve to summarize the generating-by-generalizing argument). It might be noted that this continues the long tradition seen in the creation myths found in various cultures: they inevitably star the same types of people which tell the stories. This ethnocentrism may also affect our scientific subculture: in our "creation myth", the starring roles are typically played by the clever, the technologically innovative, the communicative—indeed, by those who we might like to consider the forerunners of the first scientists.

Undoubtedly humans are the outgrowth of a whole series of selection pressures: each Ice Age has probably left behind a residue of abilities that, like the boulders scattered across the landscape, remain after the vicissitudes have vanished. And it has raised such questions as: Did the generalized animal develop out of accumulated leftover specializations—just as the general-purpose computer developed a few decades ago out of such leftover special-purpose computers as the player piano and antiaircraft aiming systems? Was there a "critical cerebral mass" beyond which there was a self sustaining selection for more brain complexity?

Such reasoning about how general-purpose humans have escaped the overspecialized dead ends of evolution may have merit, though is perhaps more relevant to later aspects of hominid development than to the origins of the great hominid encephalization:

generalization notwithstanding, some skills are more fundamental or more exposed than others, some tracks are fast tracks. Consider these summary statements:

Precision throwing has an immediate use for redundant neural timing circuits, and thus for encephalization trends that provide them.

Throwing specializations would seem a reasonable candidate for an early special-purpose invention whose neural machinery could have important secondary uses such as tool-sharpening and fire-starting.

Sequencing has major exposure to selection in that precision throwing has direct effects on food supply and fetus/infant mortality (throwing also affects intraspecies defense and aggression capabilities, but they do not have the everyday exposure to selection pressures which small game hunting provides).

Success in throwing could lead to a remarkable expansion of population into the widespread habitats of small mammals and birds; this could set the stage for the isolated subpopulations in which speciation more readily develops.

Bigger-is-faster-is-better means that selection for throwing success could have continued to enlarge the brain for millions of years, bootstrapping other unrelated functions: throwing is not a one-step cultural invention as is fire-starting but a visual-motor skill with a sustained growth curve.

Such strong statements usually cannot be made for other aforementioned traits (for example, generalization, socialization, altruism, kinship, bipedalism, etc.) nor for language itself: their effects on fitness, though not necessarily minor, are in many cases indirect and slowly cumulative; most seem unlikely to immediately modify habitat; some may be one-step inventions. As important as they may be to modern mankind, they may not have been central to the hominid encephalization.

We shall never know the true contributions of generalization except as a residual—after we first pare away the evolutionary contributions of precision throwing and other "fast track" candidates for shaping hominid characters. While unitary mechanistic hypotheses such as the throwing theory are often oversimplistic, they are necessary to expose the essential elements contributed by complexity.

Although identifying ancient throwing rocks as such may be no easier than finding speech fossils, further paleolithic prehistory may help indicate whether, for example, hominid habitats expanded into marginal foraging regions before rock-sharpening developed. Further comparisons of the sequential motor physiology of extant primates with their other lateralizations may, to the extent that history and hypertrophy can be separated, also help evaluate some of the predictions of this theory, for example, is there presently a hand-to-face-to-larynx gradient in accessibility suggesting a common sequencer? The stone throwing theory should, at the least, provide a good target at which to (metaphorically) throw stones, in the process of shaping a more comprehensive genes–selection–neuroanatomy–culture explanation for what put early hominids on the road to encephalization and language.

I thank Joan Lockard for encouragement and for aiding my education on human ethology, especially for suggesting the throwing interpretation of the infant-carry data and for suggesting that precision throwing might have derived from threat displays; Catherine Mateer and George Ojemann for many useful discussions on sequencing and language; Daniel Hartline and Katherine Graubard for suggestions on the quantitative aspects of throwing; and Dennis Willows, Richard Strathmann, and three anonymous reviewers for a critical reading of the manuscript. The Friday Harbor Laboratories provided facilities during the development of the throwing theory and the necessary experimental materials. An earlier abbreviated version of the throwing theory appeared in the December, 1981, informal newsletter of the International Society for Human Ethology. Partially supported by the National Institutes of Health, grant NS 04053. The author is an affiliate of the Child Development and Mental Retardation Center at the University of Washington.

APPENDIX B

Timing Sequencers as a Foundation for Language

WILLIAM H. CALVIN
University of Washington

Were we political scientists describing a state legislature, we could not be satisfied with a series of individual chapters portraying each legislator. We would want to know how they worked together in shifting coalitions, what determined the coalition membership on a particular issue, how leadership emerged. We would want to know if any shared facilities proved crucial in determining priorities, for example, the proverbial cloakroom or the Ways-and-Means Committee.

George Ojemann paints an increasingly detailed picture of language cortex—though one which, for reasons of space, cannot cover the whole landscape (the situation will, of course, only get worse: to paraphrase the historian H. F. Judson, editors will soon be in the position of the damned eighteenth-century Dutchman who made Rembrandt's "Night Watchman" smaller to fit a smaller room). For the mosaic of language cortex, the individual portraits are becoming clearer but the coalitions and committees are still largely *terra incognita* (present examples include the "arousal" of naming sites and the right brain's prosodic modulation of left brain language output). There are, however, some suggestions for shared facilities which hold the promise of explaining not only crucial aspects of present-day function but the evolution of language over the last 100,000 generations of the hominid lineage.

The obvious candidates for the antecedents of language specializations would be those for species-specific vocalizations (wrong location: see Sutton, 1979; also reviewed in Calvin and Ojemann, 1980) and gestures (see Hewes' 1976 review of two centuries of gesture *qua* language). Doreen Kimura (1976) suggested that perhaps a lateralization for motor sequencing could have provided a more fundamental foundation. Here I wish to interject some of the evolutionary and biophysical arguments for why precision timing and sequencing may be a yet more primitive scaffolding upon which both sign and spoken language could have been built.

Some buffer memories need not preserve sequential order, for example, a grocery list, and indeed most neurobiology emphasizes gestalt-type pattern recognition and memory. Sign language may express several sentence elements simultaneously; it does not have a prominent syntax (for example, Patterson and Linden, 1981), an ordering of elements which is important for proper decoding of meaning. But the extensive use of word-order in spoken language implies the need for a large buffer memory which preserves sequence, for example, for the recursive application of word-order rules when analyzing a sentence with several embedded phrases. On a shorter time scale, such a sequential buffer is needed to decode phoneme-order into words; indeed, to distinguish one phoneme from a closely related one may require that the neural templates underlying "categorical perception" be capable of resolving millisecond differences in a long sound sequence.

Adaptation may, of course, have gradually shaped such neural machinery because of natural selection for communications skills. But there are prominent sidesteps in evolution too, where environmental selection for one trait (feathers to keep reptiles warm is the standard example) may result in quite novel capabilities (such as bird flight) when a threshold is finally crossed (some feathers don't buy you some flight). What hominid trait that requires a precision sequential timing buffer is strongly exposed to natural selection? GET SET and THROW. To throw a stone requires that you rapidly pump out a series of motor commands (to "play a motor tape"), presumably loaded into a sequential buffer during GET SET.

The sequence required for throwing needs very precise timing, and the required precision escalates as throwing speed and distance increase. To hit a rabbit-sized target at 4 m requires that the hand release its grip on the stone within a "launch window" of about 5 ms near the top of the throwing arc. Newtonian physics says, however, that to double the target distance to 8 m means the brain and hand must coordinate their act to release the stone within a launch window eight-fold narrower. To attain 0.7 ms precision over several hundred ms of throwing time is not within the known timing capabilities of single neurons, but it can be accomplished by sufficient redundancy: to double the distance at which you can hit a rabbit could, for example, be done by 64-fold redundancy of timing circuits over whatever sufficed for 4 m throws (just as with standard deviation, this precision has a square root relation to number of samples: Calvin, 1982). A power-law exponent of six creates a considerable need for many more timers if throwing distance or speed is to modestly increase.

A brain good at temporarily borrowing such timing circuits from elsewhere in the brain ("concentrating" during GET SET serving to create a parallel circuit of various areas?), or a bigger brain having more timing circuits to start with, would be a better brain for hunting (Calvin, 1982). Throwing suggests a way in which environmental selection for action-at-a-distance predation could have bootstrapped language (not to mention handedness and bigger brains).

Besides the obvious usefulness of such a precision sequential buffer for both receptive and expressive language processing in the off-hours, when not being used for throwing, there are some reasons for thinking that language cortex indeed strongly overlaps such a sequential buffer. The interactions between finger-tapping and speech (Lomas and Kimura, 1976) and the manual-brachial sequencing deficits in aphasics (Kimura, 1976, 1982) were what led to the study of bilateral oral-facial sequencing deficits in aphasics (Mateer and Kimura, 1977) and to the mapping of oral-facial sequencing in language cortex (Ojemann and Mateer, 1979). Though the manual-brachial sequencer has not yet been mapped, the motor strip representation goes from arm to hand to face to mouth to larynx, suggesting a natural order of evolutionary improvement if borrow-from-thy-neighbor was an important principle. In addition to the hemispheric differences literature on time perception (for example, Beaver and Chiarello, 1974; Tallal and Newcombe, 1978; Vroon, Timmers, and Tempelaars, 1977), there is also auditory evidence that temporal sequence is represented in anterior temporal lobe of the dominant hemisphere (for example, Efron, 1963; Sherwin and Efron, 1980).

The 86% overlap in motor mimicry and phoneme decoding illustrates the principle of shared-use which, I suspect, will also begin to apply to the rest of the patchwork quilt of the language cortex as time and ingenuity permit more language and motor tasks to be tested in the operating room. Redundancy, a concept closely related to shared use, has also appeared in two senses: the backup system type seen in Ojemann's tumor patient where Broca's area moved across the Sylvian fissure to superior temporal gyrus, and the parallel redundancy for precision timing which is an emergent property of some neural circuits. I would suggest, however, that neither backup redundancy nor temporarily paralleled timing circuits is as fundamental a concept here as that of the coalition of individual areas, themselves having multiple functional attributes.

As we begin to look at lateral language cortex as a shifting coalition of specialized areas working in tandem for some purposes and separately for others, we may eventually perceive how leadership emerges and engages the selective attention system for coupling language planning to the motor system, for coupling thought to action.

Bibliography

The best effect of any book is that it excites the reader to self-activity *Thomas Carlyle*

General References

Barash, David. *The Whisperings Within: Evolution and the Origin of Human Nature.* New York: Harper and Row (1979). An introduction to sociobiology, highly recommended.

Blakemore, Colin. *Mechanics of the Mind.* Cambridge: Cambridge University Press (1977). In the valuable British tradition of explaining science to laymen, it is especially good on developmental aspects such as nature vs. nurture.

Calvin, William H., and George A. Ojemann, *Inside the Brain: Mapping the Cortex, Exploring the Neuron.* New York: New American Library (1980).

Darlington, Philip J., Jr., *Evolution for Naturalists: The Simple Principles and the Complex Reality.* New York: Wiley (1980). A succinct summary of the major arguments in evolutionary theory for the past century; Professor Darlington also introduces throwing's effects on brain evolution.

Dawkins, Richard. *The Extended Phenotype.* San Francisco: Freeman (1982). The selfish gene, primarily aimed at professional biologists.

Gribbin, John, and Jeremy Cherfas. *The Monkey Puzzle: Reshaping the Evolutionary Tree.* New York: Pantheon (1982). A review of the molecular clock evidence which suggests that we and the chimpanzees and gorillas separated to go our own ways a mere 4.5 million years ago.

Hubel, David H., ed. *The Brain.* San Francisco: Freeman (1979). Also a special issue of *Scientific American* 241(3) (September 1979).

Konner, Melvin. *The Tangled Wing: Biological Constraints on the Human*

Spirit. New York: Holt, Rinehart, & Winston (1982). Written by a biological anthropologist, it is a particularly good review of the nature-nurture argument and the neurobiology of behavior. It is unique for combining a broadly competent biological perspective with the philosopher's traditional concerns for emotions, ethics, and aesthetics.

Maynard Smith, John, ed. *Evolution Now: A Century after Darwin.* San Francisco: Freeman (1982). This reprints a series of recent research papers on evolutionary biology for the serious student.

Passingham, Richard. *The Human Primate.* San Francisco: Freeman (1982). He poses the question: what would it take to make a chimpanzee human?

Preface

Blinkov, Samuil M., and Il'ya I. Glezer. *The Human Brain in Figures and Tables. A Quantitative Handbook.* New York: Basic Books (1968, translated from the Russian in 1968).

Rockel, A. J., R. W. Hiorns, and T. P. S. Powell. "Numbers of neurons through full depth of neocortex." *Proceedings of the Anatomical Society of Great Britain and Ireland* 118:371 (1974). These authors would raise the 20 billion estimate for human cortex to about 50 billion.

1 | The Throwing Madonna

Annett, Marian. "A classification of hand preference by association analysis." *British Journal of Psychology* 61:303–321 (1970).

Calvin William H. "Did throwing stones shape hominid brain evolution?" *Ethology and Sociobiology* 3:115–124 (1982). See Appendix A.

Dahlberg, Frances, ed. *Woman the Gatherer.* New Haven: Yale University Press (1981). Includes a discussion of the Agta of the Philippines, where the women hunt systematically.

Goodall, Jane van Lawick. "The behaviour of free-living chimpanzees in the Gombe Stream Preserve." *Animal Behaviour Monographs* 1(3):161–311 (1968). See p. 203 for throwing data.

Lee, Richard B. "What hunters do for a living, or how to make out on scarce resources." Pages 30–48 in *Man the Hunter.* Edited by R. B. Lee and I. DeVore. Chicago: Aldine (1968).

Lockard, Joan S., Paul C. Daley, and Virginia M. Gunderson. "Maternal and paternal differences in infant carry: U.S. and African data." *American Naturalist* 113:235–246 (1979).

Lovejoy, C. Owen. "The origin of man." *Science* 211:341–350 (23 January 1981).

Oakley, Kenneth P. "Skill as a human possession." In *Perspectives on Human Evolution.* Edited by S. L. Washburn and P. Dolhinow. 2:14–50. New York: Holt, Rinehart & Winston (1972).

Salk, Lee. "The role of the heartbeat in the relations between mother and infant." *Scientific American* 228(5):24–29 (1973).

Schmidt-Besserat, Denise. "Decipherment of the earliest tablets." *Science* 211:283–285 (1981).

Woolley, L. *The Beginnings of Civilization.* New York: New American Library (1963). To this chapter's brief history of writing should be added Woolley's following proviso on sounds: "Pictorial representation ends and true writing begins at the moment when an indubitable linguistic element first comes in, and that can only happen when signs have acquired a phonetic value. The gap which divides the pictogram from the hieroglyph and ultimately from the phonetic sign is so great that for most peoples it has proved impassable. It is to the credit of the Sumarians that they were able to bridge that gap. . . . [Their neighbors then invented their own scripts but were indebted to the Sumarians] for the basic conception that a written sign might represent not a thing but a sound."

2 | The Lovable Cat: Mimicry Strikes Again

Eliot, T. S. "Old Possum's Book of Practical Cats." In *The Complete Poems and Plays 1909–1950.* New York: Harcourt (1952).

Fox, Michael W. *Understanding Your Cat.* New York: Coward, McCann & Geoghegan (1974).

Hiam, Alexander W. "Airborne models and flying mimics." *Natural History* 91(4):42–49 (April 1982).

Lorenz, Konrad. "Ganzheit und Teil in der teirischen und menschlichen Gemeinschaft." 1950. Reprinted in *Studies in animal and human behavior* 2:115–195, Harvard University Press, (1971).

Moore, Bruce R., and Susan Stuttard. "Dr. Guthrie and *Felis domesticus* Or: Tripping over the cat." *Science* 205:1031–2 (1979).

Wickler, W. *Mimicry in Plants and Animals.* Translated by R. D. Martin. New York: McGraw-Hill (1974).

3 | Woman the Toolmaker?

Boesch, Christophe, and Hedwige Boesch. "Sex differences in the use of natural hammers by wild chimpanzees: a preliminary report." *Journal of Human Evolution* 10:585–593 (1981).

Goodall, Jane van Lawick. "Tool-using in Primates and Other Vertebrates." *Advances in the Study of Behaviour* 3:195–249 (1970).

Goodall, Jane. "Continuities between chimpanzee and human behavior." In *Human Origins: Louis Leakey and the East African Evidence,* volume 3 of *Perspectives on Human Evolution.* Edited by Glynn Ll. Isaac and Elizabeth R. McCown, pp. 81–96. Menlo Park, Calif.: W. A. Benjamin (1976).

Kawai, Masao. "Newly acquired precultural behavior of the natural troop of Japanese monkeys on Koshima islet." *Primates* 6:1–30 (1965).

McGrew, William. "Evolutionary implications of sex differences in chimpanzee predation and tool use." In *The Great Apes,* volume 5 of *Perspectives on Human Evolution.* Edited by David A. Hamburg and Elizabeth R. McCown, pp. 441–464. Menlo Park, Calif.: Benjamin/Cummings (1979).

Oakley, Kenneth P. *Man the Tool-maker.* Chicago: University of Chicago Press (1964).

Pfeiffer, John. *The Emergence of Man, Second Edition.* New York: Harper and Row (1972).

Savage T. S., and J. Wyman. "Observations on the external characters and habits of the Troglodytes niger (geoff) and on its organization." *Boston Journal of Nature* 5:417–443 (1843–44).

4 | Did Throwing Stones Lead to Bigger Brains?

Calvin, William H. "Did throwing stones shape hominid brain evolution?" *Ethology and Sociobiology* 3:115–124 (1982). See Appendix A.

Calvin, W. H. "A stone's throw and its launch window: Timing Precision and its Implications for Language and Hominid Brains." *Journal of Theoretical Biology* (1983, in press). See Appendix B.

Calvin, William H., and George A. Ojemann. *Inside the Brain: Mapping the Cortex, Exploring the Neuron.* New York: New American Library (1980).

Clay, John R., and Robert DeHaan. Fluctuations in interbeat interval in rhythmic heart-cell clusters." *Biophysical Journal* 28:377–389 (1979).

Clynes, Manfred, ed. *Music, Mind, and Brain.* New York: Plenum (1982).

Enright, James T. "Temporal precision in circadian systems: a reliable neuronal clock from unreliable components?" *Science* 209:1542–1544 (1980).

Kimura, Doreen. "The neural basis of language *qua* gesture." *Studies in Neurolinguistics* 2:145–156 (1976).

Mateer, Catherine, and Doreen Kimura. "Impairment of non-verbal oral movements in aphasics." *Brain and Language* 4:262–276 (1977).

McGlone, Jeannette. "Sex differences in human brain asymmetry: a critical survey." *Behavioral and Brain Sciences* 3:215–263 (June 1980). Also see follow-up comments in June 1982 issue.

Ojemann, George A. "Brain mechanisms for language: Observations during neurosurgery." In *Epilepsy: A Window to Brain Mechanisms.* Edited by Joan S. Lockard and Arthur A. Ward, Jr. pp. 243–260. New York: Raven Press (1980).

Ojemann, George A., and Catherine Mateer. "Human language cortex: localization of memory, syntax, and sequential motor-phoneme identification systems." *Science* 205:1401–1403 (28 September 1979).

5 | The Ratchets of Social Evolution

Axelrod, Robert, and William D. Hamilton. "The evolution of cooperation." *Science* 211:1390–1396 (1981).

Calvin, William H., and George A. Ojemann. *Inside the Brain.* New York: New American Library (1980). See Chapter 7 for an introduction to right-hemisphere specializations, including chimeric face tests; chapters 3 and 4 for the language specializations of the left hemisphere; Chapter 14 for the frontal lobe role in flexible strategies.

Fried, Itzhak, Catherine Mateer, George Ojemann, Richard Wohns, and Paul Fedio. "Organization of visuospatial functions in human cortex: evidence from electrical stimulation." *Brain* 105:349–371 (1982).

Lockard, Joan S. "Studies of human social signals; theory, methods, and data." In *Evolution of Human Social Behavior.* Edited by J. S. Lockard. Amsterdam: Elsevier (1980).

May, Robert M. "The Evolution of Cooperation." *Nature* 292:291–292 (1981).

Wilson, Edward O. *On Human Nature.* Cambridge, Mass.: Harvard University Press (1978).

6 | The Computer as Metaphor in Neurobiology

Calvin, William H. "To spike or not to spike? Controlling the neuron's rhythm, preventing the ectopic beat." In *Abnormal Nerves and Muscles as Impulse Generators.* Edited by William J. Culp and Jose Ochoa, pp. 295–321. New York: Oxford University Press (1982).

Calvin, William H., and George A. Ojemann. *Inside the Brain: Mapping the Cortex, Exploring the Neuron.* New York: New American Library (1980).

Roberts, Alan, and Brian M. H. Bush, eds. *Neurones Without Impulses.* New York: Cambridge University Press (1981).

Selverston, Allen I., et al. "Are central pattern generators understandable?" *Behavioral and Brain Sciences* 3:535–571 (1980).

Stevens, Charles F. "The neuron." *Scientific American* 241(3):55–65 (September 1979).

7 | Last Year in Jerusalem

Katchalsky, Aharon, Vernon Rowland, and Robert Blumenthal. *Dynamic Patterns of Brain Cells Assemblies.* Cambridge, Mass.: MIT Press (1974).

Werman, Robert. "Let us build a nervous system: a fragment of an alleged ancient Hebrew manuscript reputedly found in the Judean desert." *Progress in Neurobiology* 11:309–313 (1978). A neurobiological and biophysical version of the creation myth, indeed presented as a Talmudic conversation between two angels. The original version was read at the Mount Scopus conference as an entertainment.

8 | Computing Without Nerve Impulses

Bullock, Theodore Holmes. "Neuron doctrine and electrophysiology." *Science* 129:997–1002 (1959). This is the influential review which predicted the existence of nonspiking neurons: "[It would probably occasion little surprise] were someone to find a neuron which gave no all-or-none impulses but whose axon carried only graded and decrementally spreading activity. This may well be the primitive property, and it may well be retained in the many very short axoned neurons in the highest centers of both invertebrates and vertebrates."

Calvin, William H. "Normal repetitive firing and its pathophysiology." In *Epilepsy: A Window to Brain Mechanisms.* Edited by J. S. Lockard and A. A. Ward, Jr. pp. 97–121. New York: Raven Press (1980).

Calvin, William H., and Katherine Graubard. "Styles of neuronal computation." In *The Neurosciences, Fourth Study Program.* Edited by F. O. Schmitt and F. G. Worden. Cambridge, Mass.: pp. 513–524. M.I.T. Press (1979).

Graubard, Katherine. "Synaptic transmission without action potentials: input-output properties of a non-spiking neuron." *Journal of Neurophysiology* 41:1014–1015 (1978).

Graubard, Katherine, and William H. Calvin. "Presynaptic dendrites: implications of spikeless synaptic transmission and dendritic geometry." In *The Neurosciences, Fourth Study Program.* Edited by F. O. Schmitt and F. G. Worden, pp. 317–331. Cambridge, Mass.: MIT Press (1979).

Graubard, Katherine, Jonathan A. Raper, and Daniel K. Hartline. "Graded synaptic transmission between spiking neurons." *Proceedings of the National Academy of Sciences* 77:3733–3735 (1980).

Raper, Jonathan A. "Nonimpulse-mediated synaptic transmission during the generation of a cyclic motor pattern." *Science* 205:304–306 (1979).

Roberts, Alan, and Brian M. H. Bush, eds. *Neurones Without Impulses.* New York: Cambridge University Press (1981).

9 | Aplysia, the Hare of the Ocean

Alkon, Daniel L. "Voltage-dependent calcium and potassium ion conductances: A contingency mechanism for an associative learning model." *Science* 205:810–816 (24 August 1979).

Darwin, Charles. *The Voyage of the Beagle.* 1839. New York: Bantam (reprinted, 1958). *Aplysia* quote on p. 5.

Graubard, Katherine. "Serial synapses in *Aplysia*." *Journal of Neurobiology* 9:325–328 (1978).

Hawkins, Robert D., Thomas W. Abrams, Thomas J. Carew, and Eric R. Kandel. "A cellular mechanism of classical conditioning in *Aplysia:* Activity-dependent amplification of presynaptic facilitation." *Science* 219:400–405 (28 January 1983).

Kandel, Eric R. *Cellular Basis of Behavior.* San Francisco: Freeman (1976).

Kandel, Eric R. "Cellular insights into behavior and learning." *Harvey Lectures* 73:19–92 (1979).

Kandel, Eric R. *Behavioral Biology of* Aplysia. San Francisco: Freeman (1979).

Siegelbaum, Steven A., Joseph S. Camardo, and Eric R. Kandel. "Serotonin and cyclic AMP close single K+ channels in Aplysia sensory neurones." *Nature* 299:413–417 (30 September 1982).

Walters, Edgar T., and John H. Byrne. "Associative conditioning of single sensory neurons suggests a cellular mechanism for learning." *Science* 219:405–408 (28 January 1983).

10 | Left Brain, Right Brain: Science or the New Phrenology?

Bradshaw, John L. and Norman C. Nettleton. "The nature of hemispheric specialization in man." *Behavioral and Brain Sciences* 4:51–91 (1981).

Dennis, Maureen D., and Harry A. Whitaker. "Language acquisition following hemidecortication: Linguistic superiority of the left over the right hemisphere." *Brain and Language* 3:404–433 (1976).

Kolb, Bryan, and Ian Q. Whishaw. *Fundamentals of Human Neuropsychology.* San Francisco: Freeman (1980).

Sperry, Roger. "Some effects of disconnecting the cerebral hemispheres." *Science* 217:1223–1226 (24 September 1982). This is Professor Sperry's 1981 Nobel Prize lecture. He favors a different explanation for right-hemisphere language than I present here.

Wiesel, Torsten N. "Postnatal development of the visual cortex and the influence of environment." *Nature* 299:583–591 (14 October 1982). This is Professor Wiesel's 1981 Nobel Prize lecture. It shows how the cortex starts off at birth with highly specific connections to the eye, and how experience modifies them.

11 | What to Do About Tic Douloureux

Calvin, William H., John D. Loeser, and John F. Howe. "A neurophysiological theory for the pain mechanism of tic douloureux." *Pain* 3:25–41 (1977). A review of the tic puzzle and a detailed theory for a pain mechanism, involving demyelinated nerve and presynaptic inhibition steps.

Calvin, William H., and George A. Ojemann. *Inside the Brain: Mapping the Cortex, Exploring the Neuron.* New York: New American Library (1980). Better suited than the other references for the general reader. Chapters 12 and 13 have more background on neuralgias and present the multiple sclerosis version of tic douloureux.

Jannetta, Peter. "Microsurgical approach to the trigeminal nerve for tic douloureux." *Progress in Neurological Surgery* 7:180–200 (1976). Dr. Jannetta, Professor of Neurological Surgery at the University of Pittsburgh, is primarily responsible for convincing the neurosurgical world that the artery was the typical problem; today, about six other major neurosurgical centers have also accumulated substantial experience with the artery-moving operation. He has also reported that a similar artery causes damage to other cranial nerves, giving rise to hemifacial spasm, glossopharyngeal neuralgia, some types of vertigo and tinnitus, and some cases of high blood pressure.

Loeser, John D. "What to do about tic douloureux." *Journal of the American Medical Association* 239:1153–1155 (1978). A physician's guide to both the medical and surgical therapy of tic, written by my research collaborator John Loeser, Professor of Neurological Surgery at the University of Washington in Seattle, who has kindly allowed me to recycle his euphonious title; he disclaims all responsibility, however, for the doggerel verse which appeared one day, nailed to his office door. The article notes the precautions which *must* be taken when starting up drug treatment of tic: patients must have regular blood tests during their first year on carbamazepine (Tegretol is the trademarked pill), as a few individuals have slowly developed serious problems with white cell production.

Samii, M., and Peter J. Jannetta, eds. *The Cranial Nerves.* New York: Springer Verlag (1981). This book covers the arterial problems for the other cranial nerves in addition to the trigeminal.

Sweet, William H. "Treatment of facial pain by percutaneous differential thermal trigeminal rhizotomy." *Progress in Neurological Surgery* 7:153–179 (1976). Dr. Sweet, Professor Emeritus at Harvard, is one of the leading tic researchers; he and others established the radiofrequency heat lesion ("poaching" only the smaller fibers in the nerve) as a safe and effective treatment for tic, many neurosurgical centers now having substantial experience with this new needle technique which has largely replaced the alcohol injection method.

12 | The Woodrow Wilson Story

Calvin, William H., and George A. Ojemann. *Inside the Brain: Mapping the Cortex, Exploring the Neuron.* New York: New American Library

(1980). Chapter 7 puts the Wilson story in the context of other right-brain functions and disorders.
Heilman, Kenneth M., and Edward Valenstein. *Clinical Neuropsychology.* New York: Oxford University Press (1979).
Lynch, James C. "The functional organization of posterior parietal association cortex." *Behavioral and Brain Sciences* 3:485–534 (1980).
Smith, Gene. *When the Cheering Stopped.* New York: William Morrow (1964).
Tuchman, Barbara. *The Zimmermann Telegram.* New York: Random House (1966).
Weinstein, Edwin A. *Woodrow Wilson: A Medical and Psychological Biography.* Princeton: Princeton University Press (1981).
Weinstein, Edwin A. "Politics and health: the neurological history of Woodrow Wilson." *Trends in the NeuroSciences* 5(1):7–9 (January 1982).

13 | Thinking Clearly About Schizophrenia

Crow, Timothy J. "What is wrong with dopaminergic transmission in schizophrenia?" *Trends in NeuroSciences* 2:52–55 (February 1979).
Crow, Timothy J. "Two syndromes in schizophrenia?" *Trends in NeuroSciences* 5:351–354 (October 1982).
Gottesman, Irving I., James Shields, and Daniel Hanson. *Schizophrenia, the Epigenetic Puzzle.* Cambridge: Cambridge University Press (1982).
Horrobin, David. "A singular solution for schizophrenia." *New Scientist* 85:642–644 (28 February 1980).
Kety, Seymour S. "The biological substrates of abnormal mental states." *Federation Proceedings* 37:2267–2270 (1978).

14 | Of Cancer Pain, Magic Bullets, and Humor

Bonica, John J. "Cancer pain." In *Pain.* Edited by John J. Bonica. New York: Raven Press, pp. 335–362 (1980).
Calvin, William H. "Ectopic firing from damaged nerve: chemosensitivity and mechanosensitivity, afterdischarge, and crosstalk." In *Symposium on Idiopathic Low Back Pain.* Edited by A. A. White III and S. L. Gordon. St. Louis: Mosby (1982). This reviews the basic mechanisms of neuralgias in general, including the demyelination pathophysiology thought to provide the trigger in tic.

Calvin, William H., Marshall Devor, and John F. Howe. "Can neuralgias arise from minor demyelination? Spontaneous firing, mechanosensitivity, and afterdischarge from conducting axons." *Experimental Neurology* 75:755–763 (March 1982).

Calvin, William H., and George A. Ojemann. *Inside the Brain: Mapping the Cortex, Exploring the Neuron.* New York: New American Library (1980). Placebo research is discussed on pp. 103–105. See pp. 209–210 for an analysis of R and D percentages.

Cousins, Norman. "Back to Hippocrates." *Saturday Review* (February 1982):12.

Diamond, Jack, Jose Ochoa, and William J. Culp. "An introduction to abnormal nerves and muscles as impulse generators." In *Abnormal Nerves and Muscles as Impulse Generators.* Edited by William J. Culp and Jose Ochoa, pp. 3–24. New York: Oxford University Press (1982).

Eibl-Eibesfeldt, Irenaeus. *Ethology.* New York: Holt, Rinehart & Winston (1975). Includes (at pp. 87–88) a modern discussion of bird behavior to cognitive mismatch.

Levine, J. B., N. C. Gordon, R. T. Jones, and H. L. Fields. "The narcotic antagonist naloxone enhances clinical pain." *Nature* 272:826–827 (27 April 1978).

Melzack, Ronald, and Patrick D. Wall. *The Challenge of Pain.* New York: Basic Books (1983).

Melzack, Ronald, Patrick D. Wall, and Tony C. Ty. "Acute pain in an emergency clinic: Latency of onset and descriptor patterns related to different injuries." *Pain* 14:33–43 (1982). Of patients with cuts, burns, and other skin injuries, 53 percent initially felt no pain; for deep tissue injuries such as sprains and fractures, the figure was 28 percent. The majority reported pain onset within one hour, but for some the delays were nine hours or more.

Wall, Patrick D. "On the relation of injury to pain. The John J. Bonica Lecture." *Pain* 6:253–264 (1979). The quotation has been shortened and slightly paraphrased. His reference to the lassitude following hepatitis on p. 263 was made with the incident described in Chapter 7 fresh in mind; fortunately, he also arrived in Jerusalem several days too late for the Thanksgiving feast.

Young, John Z. *Doubt and Certainty in Science: A Biologist's Reflections on the Brain.* New York: Oxford University Press (1960). An expanded edition of the 1950 Reith lectures given over the BBC.

15 | Linguistics and the Brain's Buffer

Fromkin, Victoria, and Robert Rodman. *An Introduction to Language.* 2d ed. New York: Holt, Rinehart & Winston (1978).

Hubel, David H. "Exploration of the primary visual cortex, 1955–1978." *Nature* 299:515–524 (7 October 1982). This Nobel Prize lecture is a very readable story of how nerve cell connections form "templates" for visual patterns and how a hierarchy of them leads to successive levels of abstraction.

Lumsden, Charles J., and Edward O. Wilson. *Genes, Mind, and Culture.* Cambridge, Mass.: Harvard University Press (1981). Their figure on p. 59 is a useful summary of information-processing buffers.

Patterson, Francine, and Eugene Linden. *The Education of Koko.* New York: Holt, Rinehart & Winston (1981). Sign language taught to a gorilla.

Smith, Neil, and Deirdre Wilson. *Modern Linguistics: The Results of Chomsky's Revolution.* Bloomington: Indiana University Press (1979).

16 | Probing Language Cortex: The Second Wave

Calvin, William H., and George A. Ojemann. *Inside the Brain: Mapping the Cortex, Exploring the Neuron.* New York: New American Library (1980). Chapters 3 and 4 cover language physiology; Chapter 5 covers part of the thalamic story, chapters 1 and 15 cover the epilepsy operation itself.

Calvin, William H., George A. Ojemann, and Arthur A. Ward, Jr. "Human cortical neurons in epileptogenic foci: Comparison of interictal firing patterns to those of 'epileptic' neurons in animals." *Electroencephalography and Clinical Neurophysiology* 34:337–351 (April 1973).

Calvin, William H., and Charles F. Stevens. "Synaptic noise and other sources of randomness in motoneuron interspike intervals." *Journal of Neurophysiology* 31:574–587 (July 1968).

Efron, Robert. "Temporal perception, aphasia, and *deja vu.*" *Brain* 86:403–424 (1963).

Fedio, Paul, and John Van Buren. "Memory deficits during electrical stimulation of speech cortex in conscious man." *Brain and Language* 1:29–42 (1974).

Fried, Itzhak, George A. Ojemann, and Eberhard E. Fetz. "Language-related potentials specific to human language cortex." *Science* 212:353–356 (17 April 1981).

Gould, Stephen Jay. *The Mismeasure of Man.* New York: Norton (1981).

Liberman, A., F. Cooper, D. Shankweiler, and M. Studdert-Kennedy. "Perception of the speech code." *Psychological Reviews* 74:431–461 (1967).

Mateer, Catherine, and Doreen Kimura. "Impairment of nonverbal oral movements in aphasia." *Brain and Language* 4:262–276 (1977).

Mateer, Catherine A., Samuel B. Polen, and George A. Ojemann. "Sexual variation in cortical localization of naming as determined by stimulation mapping." *Behavioral and Brain Sciences* 5:310–311 (1982).

Mateer, Catherine A., Samuel B. Polen, George A. Ojemann, and Allen R. Wyler. "Cortical localization of finger spelling and oral language: A case study." *Brain and Language* 17:46–57 (1982).

Ojemann, George A. "Brain mechanisms for language: observations during neurosurgery." In *Epilepsy: A Window to Brain Mechanisms.* Edited by Joan S. Lockard and Arthur A. Ward, Jr., pp. 243–260. New York: Raven Press (1980).

Ojemann, George A. "Interrelationships in the localization of language, memory, and motor mechanisms in human cortex and thalamus." In *New Perspectives in Cerebral Localization.* Edited by R. A. Thompson and John R. Green. New York: Raven Press (1982).

Ojemann, George A. "Brain organization for language from the perspective of electrical stimulation mapping." *Behavioral and Brain Sciences.* In press (1983).

Ojemann, George A., and Catherine Mateer. "Human language cortex: localization of memory, syntax, and sequential motor-phoneme identification systems." *Science* 205:1401–1403 (28 September 1979).

Ojemann, George A., and Harry Whitaker. "The bilingual brain." *Archives of Neurology* 35:409–412 (1978).

Penfield, Wilder, and P. Perot. "The brain's record of auditory and visual experience—a final summary and discussion." *Brain* 86:595–696 (1963).

Penfield, Wilder, and Lamar Roberts. *Speech and Brain Mechanisms.* Princeton: Princeton University Press (1959).

Sherwin, Ira, and Robert Efron. "Temporal ordering deficits following anterior temporal lobectomy." *Brain and Language* 11:195–203 (1980).

Tallal, Paula, and Joyce Schwartz. "Temporal processing, speech perception, and hemispheric asymmetry." *Trends in NeuroSciences* 3:309–311 (December 1980).

17 | The Creation Myth, Updated: A Scenario for Humankind

Barash, David. *The Whisperings Within: Evolution and the Origin of Human Nature.* New York: Harper and Row (1979).

Cronin, J. E., N. T. Boaz, C. B. Stringer, and Y. Rak. "Tempo and mode in hominid evolution." *Nature* 292:113–122 (1981).

Euler, Robert C., George J. Gumerman, Thor N. V. Karlstrom, Jeffrey S. Dean, and Richard H. Hevly. "The Colorado plateau: Cultural dynamics and paleoenvironment." *Science* 205:1089–1101 (14 September 1979). An example of neolithic population boom-and-bust in a marginal, fluctuating environment. Add to this picture the more isolating environmental vicissitudes of the Ice Ages every 100,000 years, and you will have a notion of the way natural selection affected hominids with minimal technology, especially those on the fringes of the more favored tropical populations.

Goodall, Jane van Lawick. "The behaviour of free-living chimpanzees in the Gombe Stream Preserve." *Animal Behaviour Monographs* 1(3):161–311 (1968).

Gould, Stephen Jay. *Ontogeny and Phylogeny.* Cambridge, Mass.: Harvard University Press (1977).

Gould, Stephen Jay, and Elizabeth Vrba. "Exaptation—a missing term in the science of form." *Paleobiology* 8:4–15 (1982).

Gribbin, John, and Jeremy Cherfas. *The Monkey Puzzle: Reshaping the Evolutionary Tree.* New York: Pantheon (1982).

Hamburg, David A., and Elizabeth R. McCown, eds. *The Great Apes,* volume 5 in *Perspectives on Human Evolution.* Menlo Park, Calif.: Benjamin/Cummings (1979).

Harding, Robert S. O., and Geza Teleki, eds. Omnivorous Primates: *Gathering and Hunting in Human Evolution.* New York: Columbia University Press (1981). The "hunting hypothesis" about cooperative social hunting of big game—what I irreverently called Sergeant Pepper's Big Game Hunting Band, and quite different from the solitary hunting of small game which I emphasize in chapters 1 and 4—is often thought to point up a critical factor in human evolution; the editors of this book also believe this much-discussed hypothesis to be of exaggerated importance.

Herskovits, Melville J. *Man and his Works.* New York: A. A. Knopf (1952), pp. 68–69. The Cherokee creation myth is from the unpublished field work of F. M. Olbrechts of Belgium.

Jerison, Harry. *Evolution of the Brain and Intelligence.* New York: Academic Press (1973).

Johanson, Donald, and Maitland Edey. *Lucy: The Beginnings of Humankind.* New York: Simon & Schuster (1981).

Klima, E., and Ursula Bellugi. *The Signs of Language.* Cambridge, Mass.: Harvard University Press (1979).

Konner, Melvin. *The Tangled Wing: Biological Constraints on the Human Spirit.* New York; Holt, Rinehart & Winston (1982).

Lovejoy, C. Owen. "The origin of man." *Science* 211:341–350 (23 January 1981). See Johanson and Edey, Chapter 16, for more on Lovejoy's use of K-selection and locomotion considerations.

Lumsden, Charles J., and Edward O. Wilson. *Genes, Mind, and Culture.* Cambridge, Mass.: Harvard University Press (1981).

Mayr, Ernst. *The Growth of Biological Thought: Diversity, Evolution, and Inheritance.* Cambridge, Mass.: Harvard University Press (1982).

Montagu, Ashley. "Toolmaking, hunting, and the origin of language." *Annuals of the New York Academy of Sciences* 280:266–274 (1976).

Ostrom, J. H. "*Archaeopteryx* and the origin of flight." *Quarterly Review of Biology* 49:27–47 (1974). Konner (1982) tells an interesting story from his student days, about visiting one of the experts on this earliest tetrapod with feathered wings thought to be the link between reptiles and birds: "What he finally said was that he thought archaeopteryx was very much like people. This of course puzzled me, as it was calculated to do, and when I pressed him to explain, he said: 'Well, you know, it's such a transitional creature. It's a piss-poor reptile, and it's not very much of a bird.'"

Parker, Sue Taylor, and Kathleen Rita Gibson. "A developmental model for the evolution of language and intelligence in early hominids." *Behavioral and Brain Sciences* 2:367–408 (1979). See also follow-up comments in June 1982 issue.

Passingham, Richard E. *The Human Primate.* New York: Oxford University Press (1982).

Sarnat, Harvey B., and Martin G. Netsky, *Evolution of the Nervous System.* 2d ed. New York: Oxford University Press (1981).

Silman, R. E., R. M. Leone, R. J. L. Hooper, and M. A. Preece. "Melatonin, the pineal gland, and human puberty." *Nature* 282:301–303 (1979).

Tanner, James M. *Foetus into Man: Physical Growth from Conception to Maturity.* Cambridge, Mass.: Harvard University Press (1978).

Teleki, Geza. "The omnivorous chimp." *Scientific American* 228(1):32–42 (1973).

de Waal, Frans. *Chimpanzee Politics.* Harper and Row (1983). Some evidence suggesting that chimp social life is comparable to that of humans.

Young, John Z. "The pineal gland." *Philosophy* 48:70–74 (1973).

Young, John Z. *The Life of Vertebrates.* 3d ed. New York: Oxford University Press (1981). See Chapter 24, "The Origin of Man."

References—Appendix A

Annett, M. A classification of hand preference by association analysis. *Brit. J. Psychol.* 61:303–321 (1970).

Boesch, C., Boesch, H. Sex differences in the use of natural hammers by wild chimpanzees: a preliminary report. *J. Human Evol.* 10:585–593 (1981).

Boulding, K. *Ecodynamics: A New Theory of Societal Evolution.* Beverly Hills, CA: Sage, 1978.

Bradshaw, J. L., Nettleton, N. C. The nature of hemispheric specialization in man. *Behav. Brain Sci.* 4:51–91 (1981).

Calvin, W. H. "A stone's throw and its launch window: timing precision and its implications for language and hominid brains." *Journal of Theoretical Biology* (1983, in press).

—— *The Throwing Madonna: From Nervous Cells to Hominid Brains* New York: McGraw-Hill, 1983.

——, Ojemann, G. A. *Inside the Brain: Mapping the Cortex, Exploring the Neuron.* New York: New American Library, 1980.

——, and Stevens, C. F. Synaptic noise and other sources of randomness in motoneuron interspike interrvals. *J. Neurophysiol.* 31:574–587 (1968).

Cavalli-Sforza, L., Feldman, M. Cultural Transmission: A Quantitative Approach. Princeton: Princeton University Press. (1981).

Cronin, J. E., Boaz, N. T., Stringer, C. B., Rak, Y. Tempo and mode in homonid evolution. *Nature* 292:113–122 (1981).

Darlington, P. J., Jr. Group selection, altruism, reinforcement, and throwing in human evolution. *Proc. Natl. Acad. Sci. U.S.A.* 72:3748–3752 (1975).

Efron, R. Temporal perception, aphasia, and *deja vu. Brain* 86:403–424 (1963).

Enright, J. T. Temporal precision in circadian systems: a reliable neuronal clock from unreliable components? *Science* 209:1542–154 (1980).

Fried, I., Mateer, C., Ojemann, G., Wohns, R., Fedio, P. Organization of visuospatial functions in human cortex: evidence from electrical stimulation. *Brain* 105:349–371 (1982).

Glick, S. D., Ross, D. A. Lateralization of function in rat brain. *Trends in NeuroSciences* 4:196–199 (1981).

Goodall, J., Bandora, A., Bergmann, E., Busse, C., Matama, H., Mpongo, E., Pierce, A., Riss, D. Intercommunity interactions in the chimpanzee population of the Gombe National Park. In: *The Great Apes,* Volume 5 Perspectives on Human Evolution, D. A. Hamburg and E. R. McCown (Eds.). Menlo Park, California: Benjamin/Cummings (1979), pp. 13–54.

Gould, S. J. *Ontogeny and Phylogeny.* Cambridge, MA: Harvard University Press, 1977.

——, Vrba, E. S. Exaptation—a missing term in the science of form. *Paleobiology* 8:14–15 (1982).

Hall, K. R. L. Tool-using performances as indicators of behavioural adaptability. *Current Anthropology* 4:479–494 (1963).

Hamilton, W. J., III. *Life's Color Code.* New York: McGraw-Hill, 1973.

Hewes, G. W. The current status of the gestural theory of language origins. *Ann. N. Y. Acad. Sci.* 280:482–504 (1976).

Holloway, R. L. Paleoneurological evidence for language origins. *Ann. N. Y. Acad. Sci.* 280:330–348 (1976).

Kimura, D. The neural basis of language *qua* gesture. *Studies in Neurolinguistics* 2:145–156 (1976).

—— Neuromotor mechanisms in the evolution of human communication. In *Neurobiology of Social Communication in Primates,* D. H. Steklis and M. J. Raleigh (Eds.), New York: Academic Press, 1979, pp. 197–219.

—— Left-hemisphere control of oral and brachial movements and their relation to communication. *Proc. Roy. Soc.* in press (1982).

Liberman, A., Cooper, F., Shankweiler, D., Studdert-Kennedy, M. Perception of the speech code. *Psychol. Rev.* 74:431–461 (1967).

Liepmann, H. Drei Aufsaetze aus dem Apraxiegebiet. Berlin: Karger. (1908).

Lockard, J. S., Daley, P. C., Gunderson, V. M. Maternal and paternal differences in infant carry: U.S. and African data. *American Naturalist* 113:235–246 (1979).

Lomas, J., Kimura, D., Intrahemispheric interaction between speaking and sequential manual activity. *Neuropsychologia* 14:23–33 (1976).

Lovejoy, C. O. The origin of man. *Science* 211:341–350 (1981).

Lumsden, C. J., Wilson, E. O. Genes, Mind, and Culture: The Coevolutionary Process. Cambridge: Harvard University Press. (1981).

Mateer, D., Kimura, D. Impairment of nonverbal oral movements in aphasia. *Brain and Language* 4:262–276 (1977).

Mayr, E., Animal Species and Evolution. Cambridge: Harvard University Press (1963).

McGlone, J. Sex differences in human brain asymmetry: a critical survey. *Behav. Brain. Sci.* 3:216–264 (1980).

McGrew, W. Evolutionary implications of sex differences in chimpanzee predation and tool use. In: *The Great Apes,* Volume 5 Perspectives on Human Evolution, David A. Hamburg and Elizabeth R. McCown (Eds.). Menlo Park, California: Benjamin/Cummings (1979), pp. 441–464.

Montagu, A. Toolmaking, hunting, and the origin of language. *Ann. N. Y. Acad. Sci.* 280:266–274 (1976).

Ojemann, G. A. Brain mechanisms for language: observations during neurosurgery. In *Epilepsy, A Window to Brain Mechanisms,* J. S. Lockard and A. A. Ward, Jr., (Eds.). New York: Raven Press, 1980, pp. 243–260.

—— Interrelationships in the localization of language, memory, and motor mechanisms in human cortex and thalamus. In: *New Perspectives in Cerebral Localization,* J. R. Green (Eds.), New York: Raven Press, 1982, pp. 157–175.

——, Mateer, C. Human language cortex: localization of memory, syntax, and sequential motor—phoneme identification systems. *Science* 205:1401–1403 (1979).

Ostrom, J. H. Archaeopteryx and the origin of flight. *Q. Rev. of Bio.* 49:27–47 (1974).

Patterson, F., Linden, E. *The Education of Koko.* New York: Holt, Rinehart, and Winston, 1981.

Salk, L. The role of the heartbeat in relations between mother and infant. *Sci. Amer.* 228(5):24–29 (1973).

Sherwin, I., Efron, R. Temporal ordering deficits following anterior temporal lobectomy. *Brain and Language* 11:195–203 (1980).

Stanley, S. M. Macroevolution: Pattern and Process. San Francisco: Freeman (1979).

Sutton, D. Mechanisms underlying vocal control in nonhuman primates. In

Neurobiology of Social Communication in Primates, H. D. Steklis and M. J. Raleigh (Eds.). New York: Academic Press, 1979, pp. 45–68.

Tallal, P., Newcombe, F. Impairment of auditory perception and language comprehension in dysphasia. *Brain and Language* 5:13–24 (1978).

Tallal, P., Schwartz, J. Temporal processing, speech perception, and hemispheric asymmetry. *Trends in NeuroSciences* 3:309–311 (1980).

Teleki, G. The omnivorous chimp. *Sci. Amer.* 228(1):32–42 (1973).

van Lawick-Goodall, J. The behaviour of free-living chimpanzees in the Gombe Stream Preserve. *Animal Behaviour Monographs* 1(3):161–311 (1968).

——*In the Shadow of Man.* Boston: Houghton Mifflin, 1971. (page references to Dell paperback edition, 1972).

Wilson, J. A., Mellon, D., Jr. Morphology and physiology of the claw closer neurons in snapping shrimp. *J. Exp. Biol.* in press (1982).

References—Appendix B

Beaver, T. G., Chiarello, R. J. Cerebral dominance in musicians and nonmusicians. *Science* 185:537–539 (1974).

Calvin, W. H. Did throwing stones shape hominid brain evolution? *Ethology and Sociobiology* 3:115–124 (1982).

Calvin, W. H. *The Throwing Madonna: From Nervous Cells to Hominid Brains.* New York: McGraw-Hill, 1983.

Calvin, W. H., Ojemann, G. A. *Inside the Brain: Mapping the Cortex, Exploring the Neuron.* New York: New American Library, 1980.

Efron, R. Temporal perception, aphasia, and *deja vu. Brain* 86:403–424 (1963).

Hewes, G. W. The current status of the gestural theory of language origins. *Ann. N.Y. Acad. Sci.* 280:330–348 (1976).

Kimura, D. The neural basis of language *qua* gesture. *Studies in Neurolinguistics* 2:145–156 (1976).

Kimura, D. Left-hemisphere control of oral and brachial movements and their relation to communication. *Proc. Roy. Soc.* in press, 1982.

Lomas, J., Kimura, D. Intrahemispheric interaction between speaking and sequential manual activity. *Neuropsychologia* 14:23–33 (1976).

Mateer, C., Kimura, D. Impairment of nonverbal oral movements in aphasia. *Brain and Language* 4:262–276 (1977).

Ojemann, G. A., Mateer C. Human language cortex: localization of memory, syntax, and sequential motor—phoneme identification systems. *Science* 205:1401–1403 (1979).

Patterson, F., Linden, E. *The Education of Koko.* New York: Holt, Rinehart, & Winston, 1981.

Sherwin, I., Efron, R. Temporal ordering deficits following anterior temporal lobectomy. *Brain and Language* 11:195–203 (1980).

Sutton, D. Mechanisms underlying vocal control in nonhuman primate. In *Neurobiology of Social Communication in Primates,* H. D. Steklis and M. J. Raleigh (Eds.). New York: Academic Press, 1979, pp. 45–68.

Tallal, P., Newcombe, F. Impairment of auditory perception and language comprehension in dysphasia. *Brain and Language* 5:13–24 (1978).

Vroon, P. A., Timmers, H., Tempelaars, S. On the hemispheric representation of time. In *Attention and Performance VI,* S. Dornic (Ed.). Erlbaum, pp. 231–245.

Index